PADDLING THE MISSISSIPPI

One Story at a Time

Other Titles by Jonathan Wunrow & Published by *Life is Twisted Press*:

- *High Points: A Climber's Guide to Central America* (2012)
- *Adventure Inward: A Risk Taker's Book of Quotes* (2013)
- *High Points: A Climber's Guide to Central America*, Second Edition (2017)
- *High Points: A Climber's Guide to South America* (2018)
- *Never Stop Walking: A Wales Coast Path Adventure* (2021)

PADDLING THE MISSISSIPPI

One Story at a Time

JONATHAN WUNROW

Life is Twisted Press
Bloomington, IN

ISBN (Print, color edition): 978-1-7363870-3-0
ISBN (Print, black and white edition): 978-1-7363870-4-7
ISBN (eBook): 978-1-7363870-5-4

Printed in the United States of America

Cover design by Bri Bruce Productions
Front and back cover photos by Jonathan Wunrow

Life is Twisted Press
801 W. 9th Street
Bloomington, Indiana 47404
Email: jonwunrow@gmail.com

This book is dedicated to my cousin, Jeff,
who originally dreamed this dream and who had
the grit and determination to see it through.

TABLE OF CONTENTS

INTRODUCTION 1

PART 1: July 27 - September 6, 2019 15
1,022 Miles, 42 Days

PART 2: November 3 - November 7, 2019 149
69 Miles, 5 Days

PART 3: September 13 - October 26, 2020 173
1,080 Miles, 43 Days

About the Author 331

PADDLING THE MISSISSIPPI

One Story at a Time

INTRODUCTION

I like people who do, and do not chat.

- Reinhold Messner

I think it was sometime in the Spring of 2018 when my 86-year-old mom said to me during one of our weekly phone calls, "Did you know that your cousin Jeff wants to paddle the Mississippi River?" I think she'd seen something on Facebook. For years, maybe forever, my mom has been trying to get me to move away from mountain climbing and into adventures that, in her mind, were less risky. "You should call him and see if he wants some company on his trip," my mom added.

Jeff was 58 years old at the time, one year older than me, and someone who I really hadn't spent more than a few hours with, every decade or so, since we were kids. The idea of spending 90 days together on a major adventure seemed a little far-fetched. And to hear that he was planning to paddle the Mississippi River by himself seemed even crazier, since to my knowledge Jeff had never done more than a long weekend paddle in his life.

As someone who always has a running list of ten or fifteen adventures planned for the future, I realized that paddling a big river hadn't made the list. At least it wasn't on my list in 2018. I keep a copy of my adventure list in the back of my calendar, and another copy on my desk, so that I have it at the ready when my mind starts to wander, when my job gets too frustrating, or when day-to-day life starts to feel mundane. My adventure list helps me keep one foot in the realm of possibilities.

In 2014, I published *Adventure Inward: A Risk-Taker's Book of Quotes.* It's a book that helped me better understand myself, and why I've always been drawn to long-distance adventures and expeditions that often include an element of risk.

> *The ancient Greeks asked oracles for answers; Native Americans fasted in the desert; I go climbing.*
>
> \- Steph Davis

> *We must, in all activities, push to do more difficult and challenging things. But not to beat our fellows, rather to find ourselves.*
>
> \- Charles S. Houston

Once or twice a year, I pull out my adventure list and update it with new ideas, new dreams. In 2018, my list included:

- Climb Pico da Neblina (highest mountain in Brazil) – 3 weeks
- Hike the 465-mile Baekdu Daegan Trail in South Korea - 6 weeks
- Climb Huascaran (highest peak in Peru) – 4 weeks
- Ride the Trans-Siberian Railway with Leslie (my wife) – 4 weeks
- Drive motorcycles across Mongolia with Seth (my son)
- Hike the 2,180-mile Appalachian Trail - 5 months
- Hike the 670-mile Ulster Way Trail in N. Ireland with Leslie - 8 weeks
- Bicycle across Newfoundland with Leslie – 3 weeks
- Hike the 812-mile Hayduke Trail with Anthony (climbing buddy) – 9 weeks
- Climb Bellevue de L'Innini (highest peak in French Guiana) - 3 weeks
- Hike the 4,600-mile North Country National Scenic Trail – 10 months

Paddling a big river was noticeably absent from my list. I had done some paddling in my younger days, but I don't think I'd even sat in a canoe more than three times in the past 25 years. In 1982, I'd spent the summer as a canoe guide in the Boundary Waters of Northern Minnesota. And in 1989, my youngest brother, Aaron, two friends, and I completed a 41-day, 460-mile wilderness canoe trip that started in Kenora, Ontario, near the Minnesota border and continued northwest through Canada to Norway House, Manitoba. On that trip, we'd pieced together several old hand-drawn trapper's maps to create our paddling route. But other than a couple of summer family canoe trips, and some sea kayaking day trips while living in Sitka, Alaska, that was the entirety of my paddling resume.

Always up for a new adventure, I decided to follow up on my mom's tip and tracked down Jeff's email address. I sent him a brief note saying that I was interested in talking with him about his Mississippi River dream. It turns out that Jeff had been dreaming about and planning to paddle the Mississippi for years.

The Big Muddy

The word *Mississippi* itself comes from "Misi zipi," the French rendering of the Anishinaabe (Ojibwe) name for the river, Gichi-ziibi, which means "Big or Great River."

Generally acknowledged to be 2,340 miles in length, the Mississippi River touches ten states as it heads south from Minnesota to Louisiana. The mighty river goes from being 10 to 20 feet wide at its headwaters in Lake Itasca State Park, to its widest point at Lake Winnibigoshish, Minnesota, where it is wider than 11 miles.

In more recent history, the river has also been referred to as:

- The Big Muddy
- Old Man River

- The "Mighty" Mississippi
- The Gathering of Waters
- The Great River

This great river has found its way into American literature. Thanks to Mark Twain's *Huckleberry Finn* and *Life on the Mississippi*, the Mississippi River holds a place in the imagination of millions, and in the history of a nation that was built on the backbone of transporting goods and people. For many, the Mississippi has become a metaphor for American culture, conjuring images of paddle boats and steam-powered sternwheelers plying the waters from St. Louis to the Gulf of Mexico. Even the second chapter of Don Rosa's famous comic book, *The Life and Times of Scrooge McDuck*, is set on the Mississippi.

The Big Muddy has also found its way into popular music, including the iconic song "Mississippi Queen" by the band Mountain, later re-recorded by Ozzy Osbourne. Both versions are worth putting this book down right now and listening too.

Other artists who have honored this great river with a song include:

- Leon Redbone – "Mississippi River Blues"
- Talking Heads – "Take Me to the River"
- Led Zeppelin – "When the Levee Breaks"

And my all-time favorite song to listen to while I paddled down the river: "Mississippi Mud," by Hank Williams III.

Making Plans

In response to a question about why he has written so many books, Ray Bradbury said, "I'm going to quit and be dead for a long time. . . . so I'm going to do everything I can right now, while I'm still alive."

Following the same line of reasoning, climber and author

Joe Simpson said, "You gotta make decisions. You gotta keep making decisions, even if they're wrong decisions, you know. If you don't make decisions, you're stuffed."

I can't honestly say that I'd ever thought about paddling the entire Mississippi River until my mom mentioned it. But once she did, it was all I could think about for the next several months.

I decided that the best way to proceed, was to give Cousin Jeff a call to figure out whether he was someone that I could spends weeks at a time with—and, to be honest, to see if he even wanted me to tag along. Jeff had been planning this trip for years, and up until now his plans hadn't included me.

I've learned over the years that the keys to any successful and enjoyable expedition are how well the participants get along with each other and whether they share similar goals and expectations for the trip. Both of us had several questions in mind for that first call. My biggest questions for Jeff were:

- How hard do you want to work/paddle every day?
- Canoe or kayak?
- Is your goal to paddle the entire river or just a section of it?
- What types of equipment are the most important to bring?
- How much paddling have you done in the past?

As I recall, these were Jeff's questions for me:

- What kind of food do you want to bring?
- Are we allowed to bring whiskey?
- Do you have a comfortable camp chair so we can sit by the fire at night and drink whiskey?

Jeff also informed me that he had already started mapping out every town, micro-brewery, and ice cream shop within walking distance of the river. He had me at "micro-brewery."

By the end of our first call, I was sure we'd get along just fine.

Jeff knew about some of my past expeditions on Denali, Mt. Logan, Aconcagua, and hiking the Pacific Crest Trail, the Wales Coast Path, etc., and I think at the time, he was a little nervous about his lack of experience in planning and undertaking long-distance adventures. But what Jeff lacked in past experience he more than compensated for in detailed planning, training, mental toughness, determination, and enthusiasm. Once we were on the river, Jeff was more than ready.

After our initial call, my next step was to buy and read every non-fiction book about paddling the river that I could get my hands on. These included the below:

- *Mighty Miss: A Mississippi River Experience* by Gary Hoffman (2009)
- *Mississippi Solo: A River Quest* by Eddy Harris (1998)
- *Paddle Pilgrim: An Adventure of Learning and Spirit, Kayaking the Mississippi River* by Dr. David Ellingson (2013)
- *One Woman's River: A Solo Source-to-Sea Paddle on the Mighty Mississippi* by Ellen Kolbo McDonah and Susan Knopf (2016)
- *Canoeing Mississippi* by Ernest Herndon (2001)
- *Down the Mississippi* by Leo Sheridan Anderson (1992)

I also brought a copy of Mark Twain's *Life on the Mississippi* on the trip, but I honestly couldn't get through it. Too boring.

Since I'd only crossed over the Mississippi in a car, and had never paddled any of it, I wanted to be able to imagine what it would be like—imagine the wind, and the current, and the scenery. Imagine my sore shoulders and aching back. Imagine myself paddling on that mighty river. Reading the personal accounts of others who had paddled the river filled my imagination with all kinds of images.

Creating these images in advance of a big adventure helps me decide what kinds of gear and clothing I'll need to bring. It helps me get through boring days at work and gives me something to look forward to. Knowing that I have a big adventure coming up, even when it's months away, can get in the way of me being present in the moment. But the balance I've found in my life is that when I'm on an extended adventure, I'm forced to be in the moment . . . in thousands of moments for weeks at a time. Some people find this balance day-to-day, but I've always found it season-to-season, and year-to-year.

While I was reading books, Jeff was devouring online blogs of more recent through-paddlers and gaining access to paddler websites. As the far more tech-savvy of the two of us, Jeff's online sleuthing and following of others on the river ended up providing a wealth of information once we started making our way down the Mississippi.

Jeff encouraged me to buy a copy of the Army Corps of Engineer's maps for the Mississippi River. These incredibly detailed maps are published primarily for the captains of the tugboats that move huge rafts of massive barges up and down the river. They also provide information that is essential for any Mississippi River paddler. Jeff sent me electronic documents of the Army Corps maps, but I opted for purchasing the 11x17 print copies. These paper maps can be purchased from the Army Corps in two volumes: *Upper Mississippi River Navigation Charts* and *Mississippi River: Cairo, IL to the Gulf of Mexico.*

The Army Corps maps are indispensable for river through-paddlers, and definitely the most important piece of paddling gear we had. Jeff and I protected our maps like they were . . . well, like they were whiskey.

Other Key Pre-Trip Decisions

Canoe or Kayak:

Jeff – Current Designs, Vision 150 Kayak (15 feet)
Jon – Wenona, Fusion Solo Canoe (13 feet)
Training:

Jeff - Paddled his kayak almost daily for several months leading up to the trip

Jon - None

Trip Food (Freeze Dried, Dehydrated, or Grocery Store):

Grocery Store

Favorite Paddling Snacks:

Jeff – Peanut butter and honey on a flour tortilla and his mom Joanna's snack mix recipe of M&M's, Froot Loops, raisins, and pretzels

Jon – Peanut butter on bagels, Rx bars, pickles, Seth's homemade deer meat sticks, and moldy summer sausage

Whiskey:

Jeff - Jameson Irish Whiskey
Jon - Redemption Rye Bourbon

Favorite Paddling Music:

Jeff – Rush

Jon – U2, James Taylor, Keith Skooglund, and my Mississippi River playlist

Doing What It Takes

We do not stop playing because we are old. We grow old because we stop playing.

- Anonymous

I read a climbing quote one time that went, "You have to climb hard, to climb hard." The same can be said of paddling a river like the Mississippi. You have to paddle hard, to paddle hard. I remember my older brother Steve commenting about my Mississippi River trip when I met up with him and my parents as the river passed through St. Paul, Minnesota. Steve said something like, "It must be nice to just sit back in your canoe and not have to paddle, just letting the current take you down the river." To me, that's like people who say that "climbing Mt. Everest is a walk-up."

You have to paddle hard, to get anywhere on the Mississippi River. And on the days when a strong wind is blowing, you have to paddle just to stay in place and keep from going backwards. You have to paddle when towering waves thrown up by a passing tow threaten to flip you and your entire boat full of gear upside down. And when, despite how hard you paddle, it feels like you are stuck in molasses for hours on end you have to paddle even harder, to paddle hard.

In planning his trip, Jeff had known for the past couple of years that he only had five weeks to paddle from the headwaters of the Mississippi to as far as he could get in that amount of time. He'd been granted a sabbatical from his employer and had a hard stop, after which he had to return to work.

I, on the other hand, decided at the outset that I'd just keep paddling for as long as it made sense. I was planning to for sure paddle with Jeff for those initial 35 days, and then would just keep going for a while. Throughout our planning, I kept the option open to continue on the river alone, and possibly finish in one long go.

I've never liked doing trips in sections. I either want to do the entire thing, or not even start. The Pacific Crest Trail, the Wales Coast Path, and the Southwest Coast Path were all through-hikes from start to finish, and I initially wanted my Mississippi River trip to be the same. On all those adventures, most of the people we met on the trail were section-hiking. And, without fail, the people we met would say something about "wishing" they could be doing the entire trail. I know that there is some ego involved in my "need" to complete things in one continuous trip. But I think it's more the challenge of going day after day, and the feeling of accomplishment, more than ego that has me wired to do it this way.

A few years ago, I thought it might be fun to hike the Appalachian Trail in sections. I'd decided that I would hike it in one- or two-week sections over the course of several years. But, after coming home from hiking the first five days on the A.T. starting at Springer Mountain in Georgia, I knew I'd likely not go back. It just didn't feel fulfilling to do big things a little bit at a time.

Overview of Our Trip

Overall Trip Length:	90 days
Overall Paddling Distance:	2,171 miles
Number of "Zero" Days:	4
Daily Mileage Average[1]:	25.2 miles per day

Except for 12 days in the middle, Jeff and I paddled 78 days together.

Even though I started my paddle being open to the idea of through-paddling the entire length, I ended up completing the river in three parts. The reasoning behind why I ended up doing it in three parts, over a period of 15 months, is described later in the book as the decisions were being made.

Part 1:
July 27, 2019 - September 6, 2019
42 Days - Headwater to Canton, Missouri
1,022 miles Total

One zero day due to high winds & heat index of 90+

Part 2:
November 2, 2019 - November 7, 2019
5 Days - Canton, Missouri to Clarksville, Missouri
69 miles Total

Part 3:
September 13, 2020 - October 26, 2020
43 Days - Clarksville, Missouri, to the Gulf

[1] Mileage average does not include zero days.

1,080 miles Total

Three zero days: One in Vicksburg, Mississippi, and two near Greenville, Mississippi, waiting out Hurricane Delta.

The Headwaters

The generally accepted start of the Mississippi River is in Itasca State Park, where the river drains out of the North Arm of Lake Itasca. The state park makes the most out of its notoriety for being the headwaters of the Mighty Mississippi, and has a wonderful visitor's center, café, and gift store just a few hundred yards from the official start. About a half-million people visit Itasca State Park each year, and virtually all of them walk the wide path from the headwaters parking lot, so that they can see where the river begins and have photos taken of themselves literally walking across the Mississippi River at its humble beginning.

It was especially cool for me to start my adventure at the headwaters in July of 2019 because my youngest brother Aaron, a Minnesota State Park employee since he finished graduate school, had just been named the Director of Itasca State Park. He was incognito (not wearing his uniform) when he helped us carry our gear, food, and boats from the parking lot to the ankle-deep water that was the start of our trip. Our supportive wives, Leslie (my wife) and Chris (Jeff's wife) were there to see us off, along with my sister-in-law Mary, as we literally had to walk our boats through the first several hundred yards of shallow river because they were too heavy to float with us inside.

Route Selection

There is more to route selection on the Mississippi than on a typical river trip. You'd think it would be a no-brainer. Put your boat in the water, point the bow downstream, and let 'er go. But there are several situations over the course of the 2,171-mile-long Mississippi where that paddler needs to make route decisions.

Headwaters – There were several times during the first few days of paddling when finding both the correct route and a designated canoe campsite was more difficult than we anticipated.

Dodging Rocks and Strainers - The first week on the river, especially in lower, late season water, there were hundreds of barely submerged rocks covered in metal and paint streaks from previous encounters with boats. We also encountered lots of downed trees and branches poking just above or below the waterline, waiting to be smacked into.

River Right? River Left? – When the river gets wide—1,000 feet wide, half-mile wide, one-mile wide, or when the river becomes a huge lake (Winnibigoshish, Lake Pepin)—a lot of thought goes into which side of the river to be on, especially when its windy and wavey. Crossing over to the other side of a wide river or lake when the waves have built to three or four feet high is nerve-racking at best, and deadly at worst.

Avoiding Barges – When your boat is heading downriver at a few mile-per-hour clip, and a gigantic tug pushing 48 loaded barges is coming upriver at a mile an hour or so, the distance closes quickly. And getting too close or ending up pinned between a barge and the shore, can be bad. Decisions need to be made upriver as far in advance as possible, which often involves guessing what the tugboat driver is going to do.

The End of the River – Jeff and I had originally planned to paddle through New Orleans and out into the gulf to "Mile 0" as noted on the Army Corp map. But early on, several people suggested that we consider the "Atchafalaya Finish." Finishing the Mississippi River by actually cutting off the river and paddling the Atchafalaya River to the Gulf of Mexico is becoming a popular option for through-paddlers, in part because it avoids the hundreds of ocean-going barges that ply the final hundred miles or so of river in and around Baton Rouge and New Orleans.

Finishing the river at Mile 0 also involves needing to arrange for a fishing boat to come out to pick you up, since that endpoint is 20 or so miles beyond the nearest pick-up point. We opted to take the Hog Bijou option, which involves turning off the Atchafalaya at Six-Mile Lake (a.k.a. Grand Lake, a.k.a. Yellow Bijou). And then paddling, into Wax Lake Channel, along Wax Lake Pass, across the Intercoastal Waterway that heads east and west, and then, finally, taking a right turn onto the calm and winding Hog Bijou that drains into the Gulf of Mexico.

Final Thoughts Before the Adventure Begins

The Universe is made of stories, not atoms.

- Muriel Rukeyzer

In deciding to publish this book, I had to make decisions about whether to use actual names and identifying information for the people we met along the way. I also had to decide whether to share my journal entries as I wrote them at the time, or to edit out the swear words I used and "colorful" language from some of the people that we encountered. Since this book is literally the daily journal that I kept on the trip, it is a record of what I was thinking and feeling at every point along the river.

Back when my son Seth and I lived on a boat for several years in Sitka, Alaska, we had a live-aboard neighbor named Doug, whose boat was tied off a few slips from ours. Doug was a tugboat captain, who would be gone for weeks at a time, moving barges up and down the coast of Southeast Alaska. One time, while we were walking along the dock where our boats were tied off, I was explaining to Doug why I'd decided to paint our wooden live-aboard purple when virtually all of the other boats in the harbor were painted either white or gray. Doug's reply, "It's your movie," has stuck with me ever since.

Consequently, as a rule, I have decided to follow Doug's advice and include my journal in its entirety, and not delete or add to it in any way. Sorry, Mom. I did make an exception to my rule for an individual or two, that I wrote a lot about, and have a level of respect for, who I didn't want to offend in any way.

Finally . . .

When Jeff and I had that first phone call, I had my doubts about his willingness to put up with the rigor and sacrifice and fear and boredom that an expedition like this one would invariably bring. What I came to find out about my cousin Jeff is that he is one bad-ass paddler, and a lifelong friend.

For a time I play catch while the children sing;
Then it is my turn.
Playing like this, time slips away.
Passersby point and laugh, asking,
"What is the reason for such foolishness?"
I only bow.
Even if I answered, they would not understand.
Look around! There is nothing else but this.

- Ryokan

PART ONE:
42 Days

July 27, 2019 - September 6, 2019
1,022 Miles

Mississippi Creek
July 27, 2019 - Day 1
Lake Itasca Headwaters to Coffee Pot Landing – 15.4 miles

What a day!

Now that it's 10:30pm, I can safely say it was a great day. A few hours ago, like 3pm or 4pm this afternoon, I wasn't feeling so positive. After mile 10, I was sort of over the whole Mississippi River thing for today. Ten miles would have been enough for day one. That last five and a half miles wasn't necessary, based on the fact that it sucked so bad.

I was up at 6am, ready and raring to go. Of course, Leslie, Aaron, Mary, Jeff, and Chris didn't stir until 7:30am or so. I caught up on all my work emails and started the process of clearing my mental space so I could focus on the river.

Aaron bought microwave breakfast sandwiches, I made coffee, and we were packed and out the door and driving to the parking lot at the headwaters before 9am. I'd originally wanted to head out around 8-ish but in talking with Jeff last night, he thought we could cover the 15 miles to Coffee Pot Landing in five or six hours. I wasn't optimistic after yesterday's reconnaissance where we drove to the first three places where a road crosses over the river, and from what I could see, the water was super low and clogged with vegetation. Jeff suggested we leave around 10am, so we compromised at 9am. My thinking was that the later we leave the more tourists will be milling around the headwaters, asking questions. Not my cup of tea.

I was out at the car anxious to leave a little before everyone else was ready. I felt really ready to just get moving.

From the park's Headwaters Visitor Center parking area, it is about an 800-foot walk to the iconic and much-photographed headwaters of the Mississippi, the spot where the river departs Lake Itasca, and begins its 2,400-mile journey to the Gulf of Mexico.

Our loving wives helped carry our paddles, life jackets, scant gear, and boats to the official start. Today we are paddling with empty boats, and just the things we'll need for the day. It is late in the year in terms of beginning this trip. The water is very low, and the first three days are supposed to include lots of shallow water and having to get out and drag our boats. So, we thought the emptier our boats, the better for today at least.

I'm carrying my lunch, snacks, a gallon and a half of water, bug dope, bug head net, sunscreen, rain jacket, Advil, hat, sunglasses, life jacket, two paddles, and that's about it. Aaron and Mary agreed to transport all our other gear and food to our first night's take-out point at Coffee Pot Landing, which is awesome!!

A Word About Bob

Day 1, Continued

Yesterday, when we were driving around the park and checking out river depths, we stopped several times so that I could jump out and take a few photos of the river we'd be paddling the next day. At the first bridge crossing, three or four river miles from the headwaters, there was a guy standing on the bridge, looking down at the six-foot-wide "river" passing beneath. I yelled, "Don't jump!" as a joke.

It turns out he was waiting for his buddy to paddle by, and this guy said that his friend Bob was already about two hours late. He proceeded to tell me that Bob was a world class adventurer and Mt. Everest summitter. And, at the time, one of the oldest persons to summit Mt. Everest. Bob is 77 years old!

We subsequently met Bob a couple of hours later. He'd gotten lost twice earlier in the morning, on his first day paddling the river!! Also, quite a feat!! Both times, he was only about two miles from the start. Both times he got into some dense river vegetation, assumed he was lost, and turned around and paddled back to near the start. He'd finally given up for the day and was standing along the river talking with his entourage about what to do next.

I got closer and eavesdropped to hear Bob talk about his troubles on the water so I could glean some information so as not to make the same mistakes he'd made. When his group realized that we were planning on starting the next day, one of Bob's friends immediately asked if Bob could join us. They basically begged me to allow Bob to join us. One of the members of the group, I think it was Bob's brother-in-law, pulled me aside twice to ask me to please stay with Bob for at least the next few days. It was weird to get this request about an Everest summitter.

Well, after only one day of paddling with Bob, I could write a book about how ill-prepared he is for this trip.

Aaron and Mary – Our Heroes

July 28, 2019 - Day 2

Coffee Pot Landing to Pine Point Campsite – 18.5 miles

Total Mileage: 33.9

Jeff and I had a nice evening at Aaron and Mary's. We were both up around 6am, anxious to get going. I told Bob that we'd be back at Coffee Pot Landing between 8 and 8:30am. After an awesome breakfast of eggs, sausages, and toast, Aaron drove us back to the landing and we were putting our boats in the water at 8am. We were early, so Bob wasn't ready. But once we arrived, he kicked into gear. Mary brought a plate of eggs and sausages for Bob. She's really taken a liking to him. Her dad's name is Bob and I think that both Bob's are around the same age?

We planned to paddle 18½ miles today. Three miles further than yesterday, and the map has two multi-mile sections that say, "The river enters a large wetlands area prone to aquatic vegetation constriction." There is also a mile and a half section of Class 1 rapids, so that might be exciting.

We started out expecting that we'd be in and out of our boats a lot today, slow going. We'd also read that the section we would be paddling today would provide several opportunities for losing the "river" and getting off-track into dead ends, necessitating backtracking until you could find the river route again.

Despite the warnings, we had virtually no problems! There were two spots where I had to stop and think for a minute to decide which way to go. The rapids were fine. And we didn't have to get out of our boat at all! As a matter of fact, after a couple of hours, I wanted to get out of the boat but the entire day other than the rapids was mostly areas of fairly stagnant flow and hundreds upon hundreds of twists and turns through cattails and tall river grass. Since it was all marsh, there wasn't even anywhere we could stop and get out.

It's my butt and my back and legs that all took turns getting sore and stiff. I'd shift my position. Reposition my legs. Lean back into my seat backrest. Readjust my lower back inflatable pillow thing. My shoulders really started to ache, too. I took three Advil mid-morning and three more mid-afternoon. There were some boring times for sure, but overall, it was way, way better than we'd anticipated.

At mile 11.5, we went under a road bridge, and Jeff and I pulled over to the shore, grabbed our water and lunch snacks, and sat up on the bridge waiting for Bob. We had faint cell service, so I texted Aaron that we were making good time and asked if he could meet us at Pine Point Campsite at 5pm instead of 6pm.

The remaining seven miles just snaked through the marsh. A few times, three- to four-foot-tall reeds and wild rice completely obscured the river, so we had to paddle right through it, hoping we were still in the main flow. I enjoyed being in the lead all day and doing the route finding. But for the most part, the river, now about twenty feet wide, was easy to follow.

Around mile 15, I paddled past a wooden boat ramp and saw a bright yellow metal folding chair sitting up in the woods. I thought it would be funny if I jumped out of my boat, brought the chair to the river's edge, and was just sitting in a bright yellow folding chair when Jeff paddled by. So, in my excitement over this brilliant idea, I jumped out of my boat. The water was deeper than I thought. I lost my balance and went totally under the water. Soaked. I realized later that my sunglasses disappeared sometime during the incident.

So, when Jeff paddled by a few minutes later, there I was, dripping wet, and sitting in the bright yellow folding chair. I guess it wasn't as funny as I thought it would be.

Around mile 18, I knew we were getting close to the location of the Pine Point camp spot, but we paddled and paddled and never saw an obvious way to get anywhere close to the shore. Nor did we see an obvious camping site. We just knew we had to be close.

I saw a possible channel that was totally grown in with reeds and cattails, and pointed it out to Jeff, but decided that there was no way it was the water path to the site. I paddled another quarter mile or so and saw nothing, so I headed back upstream. Bob was floating at that little indented spot, apparently waiting for further instructions. So, I said, "Bob, I'm just going to plow through here and check it out." And, sure enough, 300 yards through thick aquatic vegetation, I got to shore and could see the Pine Point Campsite up ahead.

I yelled back to Bob and Jeff, and they followed. I stood on the shore and took a couple of photos of them coming, and all I could see was the tops of their kayak paddles above the weeds.

Aaron and Mary showed up around 4pm. We had arrived around 3:30pm. They brought cooked brats and chips and beer. It was so awesome, and great to see them at the end of a hard day. If we could only get them to keep doing this at the end of every paddling day!

We drank beer, ate brats, cracked jokes, and told stories until it started to rain. Jeff and I gathered our lose gear and scrambled into our tents while Bob struggled to set his up. The last I saw of Bob, his sleeping bag was laying all rolled out in the rain and he and Mary were running around gathering up his stuff that was strewn everywhere.

We actually had okay cell service at the campsite. I've been texting Leslie on and off for the last hour or so. She's really worried about our dog Goldie, who isn't doing very well. Jeff and I have also been texting back and forth between our tents, and totally cracking each other up. So, it was an early night. By 6:30pm, Aaron and Mary had gone, and we were in our tents. I read a little and texted and was fast asleep by 8pm.

Getting Lost, and It's Only Day 3

July 29, 2019 - Day 3
Pine Point Campsite to Bemidji, Minnesota – 23 miles
Total Mileage: 56.9

Awake at 5am after a fitful night's sleep. I was sore. Couldn't get comfortable. Hate my dinky pillow! By 5:30am I was rustling around in my tent, and once I heard Bob moving around in his, I got up. It rained well into the evening, so tent flies were soaked, along with anything still on the clothesline that I'd hung up to dry out some things.

Boiled water for coffee on my MSR Reactor stove and had a Pop-Tart and a breakfast fruit bar to start my day. We all packed up and made several trips, carrying out bags from the Pine Point tent site down to the boats. Bob had a head start, and I told him he should shove off first, but he kept dawdling around. I loaded everything in my boat and pushed off, back through the tall weeds and reeds we'd pushed through to get to shore yesterday afternoon. Jeff was a little slow going this morning. Not sure why. So, he headed out last, maybe 20 minutes after me? Bob was somewhere in between.

The next seven miles were basically through an expansive marsh. There is a river, but in countless places, the river disappears into a wall of wild rice and other aquatic weeds. By late July, this part of the river is completely overgrown and choked with plants. I had to pay constant attention to the color of the vegetation: light green means new growth, plants that probably recently grew up in the middle or edge of the river; darker green meant older plants what were likely there year-round and growing more on the edge of the river; and cattails, while still in water, were way off to the side of the main river. I was also always looking down into the water to watch the direction that plants were leaning beneath the surface. The more they leaned in one direction, the more likely I was still in the main part of the river with a little current. No underwater weed bend, no current.

It was windy all day long. Really windy. The plants above the water were bent over in all kinds of directions, and every 100 to 200 feet, the river would take a sharp turn to the right or the left, and twist around on itself. Twisting and turning, mile after mile.

It was really difficult to stay in the main river and not wander off into an oxtail or alternate water route that eventually led to nowhere. And the marsh was a mile or more wide throughout, so you could wander off course for quite a way before hitting the cattails and eventual shoreline. There were several times I had to stop and really assess my surroundings to figure out which way to go. And several times I doubted my decision until I got to open water and knew I was back in the main part of the river. A little nerve-racking.

Bob had told us the day before that he had an air horn for emergencies. He said, "If I'm ever in trouble, I'll do two blasts on the horn."

About two hours into my paddle--I'd left at about 7:15am—I hear two blasts on Bob's air horn coming from a long way behind me. Shit. I immediately got out my cell phone, saw that I had a few bars of reception, and texted Bob and Jeff, "Are you okay?" I waited for a response . . . 10, 15, 20, 25 minutes. I decided that paddling back upstream would be futile because of how confusing the route had been, and not sure I'd be able to retrace my "steps." I called and texted Jeff and Bob again. No response. I tried texting Leslie and Mary to make sure I could get a signal out, and they both responded right away. Jeff is super worried about draining his phone battery, so he typically keeps his data, and consequently his texting, turned off.

Finally, Bob texted that he was "Okay."

"Have you seen Jeff?"

Bob wrote back that he had not.

So, I continued paddling, and winding my way through the tall grass, wild rice, and occasional cattails, until I reached the Iron Bridge Campsite after about six and a half miles. I pulled out, walked up to the camp where I could get a good view of the river, and texted Jeff several more times. After 30 minutes, Bob came into view and climbed out of his boat. He was soaking wet. He apparently had gotten lost at some point and couldn't turn his boat around in the thick weeds, so he jumped out of his boat, in water over his head, and dragged his boat out backwards, into the main channel. It was windy and cloudy and chilly, so I encouraged Bob to put on some dry clothes.

We waited for another hour, all the while doing contingency planning for what to do if Jeff never showed up. What should we do? What might Jeff do? Had he gotten lost and turned back? He'd obviously gotten turned around in the vast marsh, but it was impossible to know where, and would be futile to paddle back and try to find him. The wind made yelling and shouting his name equally as futile. So, we waited.

Ninety minutes after I'd pulled off the river, Jeff finally came into view. Phew!!! He'd gotten lost. Gotten stuck. Gotten scared. But he figured his way out of it and was safe!

Our plan of getting in 21 miles today and having a cold beer in Bemidji seemed dashed, but at least Jeff and Bob were safe.

Bob and I got back into our boats. Bob pushed off first and headed downriver. I turned around to look back at the campsite and saw a pile of Bob's stuff that he'd left sitting on the picnic table. His shirt, headlamp, winter hat, and jacket were all sitting there. I grabbed all his stuff, and stashed it in my boat, shaking my head once more at his lack of attention to important details.

With all of my little ducklings back together, we paddled on. Jeff and I talked along the way about how to deal with a similar situation the next time it happens. I also talked with Bob about the fact that we are all independent paddlers on this river who happen to be paddling together right now, so we all need to be one hundred percent responsible for our own route finding, gear, and safety.

The rest of the day was really beautiful. We eventually paddled out of the marshy area and into a wider, vegetation-free section of river, passed a few farms and some private riverside homes and cottages. Our first signs of people other than the road bridges we paddled beneath.

It was overcast and breezy all day, which meant no bugs! And very little direct sunshine. I never even put on sunscreen, although I should have. We paddled into a three- or four-mile section that meandered through a broadleaf forest. The map marked the area as having "numerous log jams." There were hundreds of downed trees with the constant assessment of whether we'd have to stop and unload our over-packed boats to get over the logs. But none of us ever had to get out of our boats even once, despite the low water level. There was always just enough water and flow (just a few inches needed) to scape over the top of logs or branches. I got stuck four or five times, high-centered on a barely submerged log, but was always able to wiggle free.

There was enough of a current that I also slammed the side of my canoe into some big logs and stumps. Several times I heard my Kevlar hull make a cracking noise.

The last part of the day we paddled across Irving Lake, just before Lake Bemidji. Since we were making good time today, I stopped under a bridge to get some shelter from the sun, and reserved two rooms at Bemidji's Hampton Inn, right on the lake. That would be our endpoint for the day. Something to paddle to. To get there, we had about a two-mile crossing through the middle of Irving Lake.

Jeff and I oriented the map and figured approximately where we thought we needed to cross to hit the outlet, and then headed out into two-foot-high waves.

Well, we guessed wrong on the direction, and the waves ended up being more like three feet high. I hadn't buttoned down my custom spray deck before heading across, and two waves came over the side of my canoe, mid-crossing. Paddling across a two-mile open stretch compounds just a few degree direction miscalculation into ending up way off course.

I started getting nervous about taking on too much water and then tipping over out in the middle of this big wavey lake. It only takes a few inches of water in the bottom of this small canoe loaded with a few hundred pounds of gear and me to really get tippy. I started feeling way too uncomfortable and decided to make a beeline to the nearest point on the shore, off to the right, and even that made for 30 very nerve-racking minutes of paddling, as I surfed these big waves towards the shore.

Then it was a good hour of paddling straight into the wind to get back to the channel that led to Lake Bemidji. We were all exhausted from battling the wind. And I was a bit rattled. Two mistakes: misjudging the right direction and distance, and then attempting to go out in a windy lake without my spray deck attached.

I won't do it again.

Extra Baggage

July 30, 2019 - Day 4
Bemidji, Minnesota, to Star Island (north camp) on Cass Lake - 25 miles
Total Mileage: 81.9

We had a great paddle today.

25 miles!!

I never expected we'd get this far today. And we had our first marked portage which took almost an hour to get across.

There were lots of different types of paddling again today. First, three miles across Lake Bemidji after our 7:15am start from the hotel. Then some shallow, rocky river miles, followed by several slow and twisty river miles. Then, three miles across Allen Bay, part of Cass Lake, and two or three more miles directly through the middle of Cass Lake past two Potato Islands. We're camped on the north end of Star Island at a beautiful site with a fire pit and a nice sandy beach. We even had a campfire tonight.

All in all, we were on the water for about ten hours today. Long day!! But I'm glad we got this far.

We cooked our first dinner tonight. I made tika masala chicken over basmati rice. I made enough for all three of us to eat. Jeff busted out his Jameson Irish Whiskey for a before dinner drink. And Bob was Bob. He just tagged along. Relatively worthless to our overall effort. He adds nothing. Still carrying half of his shit in large garbage bags, despite going to a camping store last night in Bemidji. He never has any idea where we're heading. No sense of direction. He hasn't even looked at a map. He refuses to be the boat out in front because he has no idea where to go.

Bob was a disaster on the portage crossing today, and totally needed help from me and Jeff. Super nice guy but adds nothing to the effort. Oh well. I don't like to talk about Bob as if he's extra baggage. But I didn't plan this Mississippi River trip with the role of being a river guide in mind. I don't want to feel like I have to worry about him, wait for him, be responsible for him, cook for him, navigate for him, etc. But it feels irresponsible and mean to tell him we're leaving him behind. He underestimated the river and this trip and is treating it like his multiple guided Everest expeditions, complete with his own personal Sherpa, cook, and expedition leader.

Day four is in the books. We did stop at a little riverside resort to see if we could buy something cold to drink. It was very sunny and warm today. I walked up to the office/store and was told that the "resort" (a few small cabins and the office) was only for guests, and they usually don't sell anything to non-guests. But they made an exception.

Oh, brother.

A couple of crabby old people.

Other than that . . . a great day!

I had a nice long talk with Leslie tonight about her new job offer while sipping my whiskey by the fire and trying to relax my stiff back. I'd intended to just start a little fire in the fire grate to burn some garbage. But I just kept adding wood and we ended up setting our chairs around a really nice campfire until the mosquitoes drove us into the tents around 8:30pm. Bob commented that he'd never sat by a campfire before.

Lake Winni – A Big Damn Lake

July 31, 2019 - Day 5
Star Island to Cass Lake to Richard's Townsite on Lake Winnibigoshish – 22 miles;
Total Mileage: 103.9

This is one big damn lake! From the Mississippi River's entrance into Lake Winnibigoshish straight across to the river outlet on the other side is 15 miles of open water paddling. All three of us (yes, we are all still together), have heard horror stories of paddlers who, while attempting to take the shortest route across Lake Winni, were surprised by sudden winds kicking up midway across, causing epic challenges and deadly tragedies.

We heard from an old guy who has lived on the lake for 44 years that last year three paddlers headed straight across, got caught in some big waves, capsized, and drowned. All three bodies were found. Everyone we've talked with advised not to attempt to paddle straight across the middle of the lake, even on a calm day. The temptation comes from knowing that to follow the lakeshore around to the other side adds at least 15 miles of lake paddling. So, 25 lake miles instead of 15. We decided before we started today that we'd follow the south shore and stay close to the shore as long as the wind and waves would let us.

Jeff promised he'd get up a little earlier today. He's typically been the last one packed and on the water. We agreed that we'd shoot for being on the water by 7am. I heard Jeff rustling around in his tent at 5:45am, so I started rustling around, too.

I am so stiff and sore. Still haven't found a comfortable sleeping position. I've always been a side sleeper and both of my shoulders ache all night long. I toss and turn all night. I need to start taking Advil just before bed. Seems like I finally started getting into a good sleep around 3am. So . . . why agree to get up at 6am and be on the water by 7am? The wind and water seem to be calmer in the mornings. And before we could face our fears on Lake Winni, we had to start the day with a five-mile crossing of Cass Lake.

I heated water for coffee and oatmeal and was on the water and paddling away from our campsite on Star Island at about 7:15am. Bob was right behind me, and Jeff maybe 20 minutes later. No big deal.

I paddled about a mile around the corner of Star Island and was hit head-on with a stout wind. Maybe blowing 15mph? I texted Jeff that Bob and I were going to start heading across, and not wait for him, since the wind was already knocking me backwards. I didn't put my spray skirt on, or my life jacket either. Looking back, both were probably a mistake. The waves weren't as big as the other day on Lake Irving, but pretty close. Maybe three-foot swells coming from my right side. In a 13-foot open boat, those are big swells. I took a compass reading and identified a point on the opposite shore to shoot for. I yelled to Bob that we were going to head for the closest point of land, and follow the shoreline, rather than continue cutting straight across the lake. The wind and waves were too strong, and I was really getting nervous. I let him go ahead for about five minutes so that I could keep him in my sights. I was getting scared, and the paddling was exhausting. It was a nerve-racking hour-and-a-half paddle, and I was very happy to be on the far side without having taken a wave over the side of my canoe and tipping over. Next time, spray skirt and life jacket for sure. That was dumb.

Between Cass Lake and Lake Winnibigoshish, the river meanders and twists for 12 miles. It was sunny with the wind in our faces most of those 12 miles. We passed a few river cabins and eventually made it to the west end of Winni around 1pm. There were two fishing lodges. We stopped at the second one, the Four Seasons, and the staff was super nice. There was an employee at the dock selling gas for fishing boats. He told us all about the big lake and gave us lots of cautions. He directed us to the lodge, where the owner's daughter and a couple of other ladies welcomed us. We bought, and they served, frozen pizzas, so we bought one to split and hung out for 45 minutes. It was a really nice stop.

The grandpa of the lady that served us has been at the Four Seasons for 44 years. He told us to be very careful on Lake Winni, and he's the one who told us about the three paddlers who capsized and died. They'd attempted to cross earlier in the year, when the water was colder, like April or May. He was very adamant that we should stick to the shore and not attempt a direct crossing. He also said that a lot of through-paddlers on the Mississippi just skip the lake crossing and get a ride by road to the other side, and that for

$25 he'd be happy to give us a ride with all our stuff. Talk about tempting! He said that lots of paddlers don't think of the large lakes like Cass and Winnibigoshish as parts of the actual river, and therefor just skip paddling across those large and sometimes dangerous lakes.

Before we shoved off from the Four Seasons dock and out into Lake Winni, I attached my boat spray skirt and put on my life jacket. I felt a lot safer. We paddled about one mile to the lake, and looking out across it, it seemed like an ocean. In parts you can't even see the other side. Its massive. We had decided to paddle about five miles along the south shore to Richard's Townsite Boat Landing. The wind was in our faces the entire time, and it was the end of the day. My shoulders really ached. We stayed close to the shore and the wind was only about 10mph or less. So, a long, hard five to six miles, but not scary at all. Hopefully tomorrow will be the same.

The boat landing had a gravel parking lot, pit toilet, and nice grassy areas with a big sign that read, "No Camping Here." We looked around for another tent spot, but there weren't any. So, we set our tents up right next to the No Camping sign and rehearsed what we'd say if someone came to kick us out.

We'd been planning on meeting up with Aaron and Mary for dinner tonight, so Bob agreed to stay back at camp and watch our stuff while Jeff and I went to dinner. It was a very nice gesture on Bob's part.

It was awesome to see Aaron and Mary again. Our third evening with them in five days. We ate at The Big Fish restaurant near Bena, Minnesota, and were served by the owner, Al Hemme. It was a cute little mom-and-pop restaurant, popular for fishermen and summer cabin owners on Cass Lake. The Wednesday night special was fried chicken, so that's what we all got. I brought my cell phone, back-up battery and charger to get all charged up while we ate and topped off my two empty water bottles.

Back at camp, after seeing Aaron and Mary off, we sat in our camp chairs, Jeff and I going over our maps and tomorrow's route and me texting Leslie and Seth until the mosquitoes drove us into our tents around 9pm. I hope I sleep better tonight. I'll start with a couple Advil.

Winnibigoshish

August 1, 2019 - Day 6
Richard's Townsite on Lake Winni to Crazy James Point Campsite – 18.4 miles
Total Mileage: 122.4

Well, we didn't get as far as I'd hoped today. Mostly a result of the fact that after this campsite, at Crazy James Point, there isn't another designated site for 12 miles and we'd already gone over 18 today, with 13 of those 18 on Lake Winnibigoshish, which was a lot of hard and windy paddling. If there'd been a place to camp in six or seven more miles, we wouldn't have kept going. We stopped at 2:30pm, so we had plenty of time to go further. I was suggesting the option of pushing on, and finding a spot to camp along the river, but the last five miles were really weedy and boggy, with no promise we'd find anyplace to set up three tents. So, we stopped.

We're officially one day off from Jeff's original itinerary. I'd love to start whittling away at making up that day, but that would mean putting in longer days. I think that once we start passing more towns, we'll have more options for places to stop. Right now, we're pretty dependent on stopping at the few designated river campsites.

With the early afternoon stop, I was hoping to get in some office work, but there is virtually no cell service here. I was able to read and send a few intermittent emails, but that was it. So, I just journaled and relaxed.

Once we got here, and got tents up and organized, I went down to the river and washed up, and laundered some underwear, shorts, and a t-shirt. Bob and Jeff didn't do either. I actually haven't seen either of them wash up in the river or a lake yet on this trip. Imagine that. I'm the clean and hygienic one! Leslie would never believe it.

I'd like to push things a little harder and try to get in a few extra miles the next few days. I still have it in my head that I might keep going after Jeff's 30 days on the river. I'd like to get as far down the river as we can.

I started to get a little worried when Jeff's itinerary for the day after tomorrow shows 25 miles, but he's suggesting that we stop in Cohassen for lunch and then Grand Rapids for dinner, cutting short the last six miles of the day. Not a big deal, but it all adds up. I want to have fun, too, so just need to find the balance.

I just asked Jeff if he planned on finishing the Mississippi River next year, in 2020, and he basically said no. I'd been assuming, I guess, that we'd do half this year, and finish it next year. His plan is to just paddle a couple of weeks for each of the next several years to finish the river. So, with that answer, which I'd never thought to talk through until now, I'm wondering if I should go home in early September, take two months off the river, and then come back and finish in late October and November alone. That would free up next summer to spend with Leslie and do a trip with her and spend time at our Rice Lake cabin.

Back to today. . . .

We were all up and out of our tents around 6am. We're going to bed so early, like 8:30 or 9pm, so I'm wide awake by 5am and then just laying there until 6am. I had my best night's sleep of the trip (other than the first night when we slept at Aaron and Mary's). I know it's because I took Advil before I fell asleep, so I'll do that again tonight.

Coffee, Pop-Tarts, and a breakfast bar. I packed up and loaded my boat much more quickly than the other two. I'm not sure why. Bob is always a little discombobulated. And Jeff? It just takes Jeff a while in the morning.

We were all a little anxious about heading out onto Lake Winni this morning. Jeff and I decided on a point off in the distance to shoot for, on the opposite shore. We could head straight for Tamarack Point, which is eight or nine miles across the open lake. But instead, we headed to the right of the point by a few miles, so we only had to paddle across five miles of open water. Maybe six. I took off when I was packed and ready, and Bob wasn't far behind. I left around 7:10am. Within a few minutes, the wind started picking up, and after 30 minutes, occasional white caps started to form. I was all buttoned up with my spray deck and skirt and had my life jacket on. The waves hit from my right side, almost broadside to the boat. I was nervous, but just kept paddling hard and staying focused on the point ahead. Five miles of open water paddling on a big lake in wind and waves and in a very small canoe, by yourself, makes for a long damn way to go.

It seemed like the point on the shore that I had headed towards never got any closer. I started counting 500 paddle strokes before staking a short break. And then 500 more. After about two to two and a half hours, the waves seemed to be getting a little smaller. Once I got within a half-mile of the shore, I turned south and headed for the point, or what I thought was Tamarack Point. Now, the wind was at my back, which was still uncomfortable to be paddling with waves coming up behind you. They have a way of hitting the boat, and then swinging the stern one direction or another. But at least I was moving faster. The turn around the point took a lot longer than I thought. And then it was another three miles to the Lake Winni dam and portage. In all, it took almost five hours of constant paddling. I was ready for a break.

The portage crossed a road and went through a park. We hauled our boats and gear up a steep bank that was overgrown with weeds, strapped on our portage wheels, and pulled our stuff about 300 yards to the put-in, and then had lunch at a picnic shelter. After six days, I'm already tired of my lunch options. Peanut butter and jelly bagels, gorp, and PowerBars. I need to buy some cheese, chips, tuna, flour tortillas, and maybe some carrots? Just to mix things up a bit.

After lunch, we paddled five and a half miles to our campsite at Crazy James Point, and here we are.

Tomorrow we were planning on going 25 miles to a state park, but Jeff is constantly contacting "River Angels" and found one who lives 23 miles downriver who offered to let us stay at their house tomorrow night. I was a little hesitant. Sort of weirds me out. But Jeff really wants to do this. I'm worried that this River Angel thing will start coming up more regularly as an option. My only real issue (other than being a little on the reclusive side of the outgoing scale) is that it won't get us as far as we'd planned to go tomorrow, and Jeff really wants to stop in Grand Rapids for another shorter day, and I'd rather we focus on going more miles, not less, since we are only one week into this trip. Jeff said that Chris (his wife) will be really happy when she finds out that we are staying with River Angels.

It's only 6:50pm. I'm bored and ready for bed, and biting flies are gnawing at my ankles. It's fairly hot still, and the breeze has completely died down. I may need to head into the tent early just to get away from the biting flies.

I am so happy to have Cass and Winnibigoshish Lakes behind us. They are crossings that both of us have been dreading for months.

Our First River Angels - Jeff and Sandy

August 2, 2019 - Day 7
Crazy James Point to River Access #10 – 30 miles
Total Mileage: 152.4

Our biggest paddling day yet: 30 miles. It didn't seem possible those first couple of days. Our next big goal will be a 35-mile day.

Tonight, we are staying with our first "River Angels," Jeff and Sandy at River Mile 1201. Jeff (my cousin) is part of a River Angel Facebook page that includes scores of people who live along the Mississippi River and are willing to help paddlers. Jeff brought the idea up yesterday of staying with these folks, and I was a little apprehensive. Not really my kind of thing, staying at strangers' houses. Jeff is really into the River Angel concept. He's already gotten advice and suggestions about water levels, camping spots, etc., from several folks via messages on Facebook.

Our campsite at Crazy James Point Campground was super buggy, so I was motivated to get going and on the water this morning. We were all out of our tents around 6am, and then it's a cluster of packing, making coffee, eating (I had oatmeal with craisins today), hauling everything to the boats, loading them, and pushing off. This is Day #7 and I've gotten into a routine in the morning for getting all packed up. Bob's routine is to have his shit strewn all over the ground, most of it outside his dry bags and garbage sacks from the night before, and then ask us for hot water for his morning tea, and then slowly get his shit together. Jeff is very organized. But no matter when he gets up, he's always that last one loading his myriad small stuff sacks into his kayak hatches.

Super flat water this morning. No breeze. Pretty cloudy, so it was a perfect morning to paddle. A few black flies found their way into the bottom of my boat and gnawed on my ankles, but it was otherwise a perfect morning to paddle. Since I left ahead of everyone this morning, I used my lightweight canoe paddle for the first hour or so. I love using it, but it's a little slower going than when I use my double-bladed kayak paddle. Fewer strokes per minute.

The river was very snaky and winding. It is getting wider and wider, but still twists and turns, and the banks are lined with reeds and cattails. I saw a couple of eagles, startled some deer drinking water, saw some herons, and chased several small flocks of geese off the water as I paddled past them.

I tried one shortcut to cut off a bend, but after a few minutes of paddling it didn't pan out. Then around 10am I tried another one that I was sure would hook back up with the main river, but after about ten minutes of paddling, I started doubting myself and checking my location on my phone's Google maps app and realized that my "shortcut" was actually Ball Club River taking me to Ball Club Lake. Oops. I had to paddle back against the current to get back to the Mississippi. After the detour, I wasn't sure if Bob and Jeff were ahead of me or behind. I dawdled around a little, and within a few minutes, Jeff came paddling around the corner.

By noon we realized that we were making great time and had already traveled 16 miles, and that we'd be to the River Angels' house by 2pm with a whole afternoon left that we could be paddling further. Jeff suggested that we just stop and say hello, and then paddle another five or six miles to camp. I readily agreed since I was still wary about staying with complete strangers.

It was a warm and muggy day and when the sun finally came out from behind the clouds, it was really hot. I dipped my Green Bay Packer hat in the cold water several times and then put it on my head so the water could run down my face to cool me off for a few seconds.

We got to River Mile 1201 (1,201 miles above the town of Cairo, which is Mile 0 for the Upper Mississippi) and saw Jeff and Sandy's house, and Sandy was already walking across her yard towards the river to meet us. She and Jeff were super friendly and have lived along the river for most of their lives. They are two of the original River Angels and told us stories about all kinds of people they've hosted over the years. They built a bunk house for paddlers, have a trailer they offer when the bunkhouse is full, and even host paddlers in their home. Jeff immediately brought us each a cold beer as we sat on their porch and visited.

We stayed for about an hour and a half and then headed out. Jeff and Sandy mentioned that the camp spot we were shooting for tonight wasn't very nice and that they'd be happy to pick us up if we changed our minds. Once we got to the site, it only took a few minutes to realize that they were right, so we called Jeff and he came with his boat trailer and hauled us back to his house for the night. I'm so happy we called him and felt kind of dumb not just staying at their house initially. Once we got settled into the bunkhouse, we treated them to dinner in the nearby town of Deer Lake at a little mom-and-pop restaurant that was connected to a bowling alley. Burgers, pizza, pop. All yummy. Then we stocked up on groceries. I bought way more than I needed, but got stuff I've been craving, like cheese, chub, tortillas, tuna, powdered milk, and granola, as well as some more PowerBars. My food bag must be up to 50 pounds!! No more food buying for me for a couple of weeks!!

It was a great day. We got our miles in and I'm sleeping in a cozy bed with two cozy pillows. The air conditioner in the bunkhouse is blasting, all my devices are charging, clothes are laundered, and I am clean as a whistle. I couldn't ask for anything more. Maybe Jeff is right about this River Angel thing.

Thunder and Lightning, Very, Very Frightening

August 3, 2019 - Day 8
River Access Point #10 to Grand Rapids, Minnesota, to Suchens Campsite – 22 miles;
Total Mileage: 174.4

A very eventful day! With some excitement at the end!!

We were all up moving in the little bunkhouse by 6am. I slept great. Cozy mattress. I slept on a quilt with my sleeping bag on top. Once I'd killed a dozen or so mosquitoes last night, I fell sound asleep. I only woke up to pee once. We got everything packed up and loaded into the truck, and then went to the house at 7am for an amazing breakfast of sourdough and fresh blueberry pancakes, sausages, and homemade maple syrup that Jeff tapped from his maple trees. He boils down the sap to make about three gallons a year. Coffee. Juice. It was amazing.

The morning was a long, quiet, winding paddle. No breeze. Sunny. Hot. I dunked my Packer cap in the water dozens of times to stay cool. I was ahead of Jeff and Bob for the first 12 miles to Cohasset. I never saw them for three hours. I had my little transistor radio on, listening to *NPR* to keep me company and pass the hours. I stopped at a boat landing in Cohasset to stretch my legs and eat a Clif Bar dipped in peanut butter. Bob eventually caught up, and we paddled together for a short bit to a short 45-yard portage around a dam that led to a small state park campground.

I'm pretty set on each of us being able to portage our own gear and boats. For example, if we end up splitting up, or Bob goes out on his own, he needs to be able haul his own shit. So, I left him on his own to transport all of his stuff today, since on the past portages Jeff and I have offered to help carry some of his gear.

I filled my water bottles at the park and headed out to paddle the next three miles to a 1,200-yard portage around a paper mill and dam entering Grand Rapids.

The portage was well-marked, but the trail was completely overgrown. Just seemed like not many people are on the river this year. Jeff's friend John met us. We had to haul our gear and boats a couple hundred yards to his truck but then he drove us around to the other side of the dam to the put-in site. By the time Jeff and Bob arrived, I'd hauled my stuff up the overgrown path through what appeared to be someone's backyard and out to the road, and I helped Bob haul his stuff up so we could speed things up a bit.

The city of Grand Rapids has built a small paddler's campsite next to the public library right along the river. We unloaded all of our gear there. I stayed with our stuff and had a nice long phone call, first with Seth and then with Leslie, while the other three went to an Army/Navy store to buy a dry bag for me, and Bob shopped for a new headlamp and dry bags to replace his garbage bags (he'd left his head lamp floating in the bottom of his boat a few days ago). One of my two big "waterproof" food dry bags has several holes eaten into it by a hungry raccoon a few nights ago.

The gang stopped at Subway and picked up some sandwiches for everyone and brought them back to the riverside campsite and picnic table so we could all eat together. It was 3pm and still super sunny and hot, so we stayed another hour just hanging around and slowly packing up in the shade. I put my 40 pounds of food into my new dry bag and ditched the one that the raccoon destroyed.

We took off around 4pm for the final six-and-a-half-mile paddle to Suchens Campsite. Bob took off first, which was promising, and Jeff and I followed 10 or 15 minutes later. After about 45 minutes, a wind picked up and blew pretty hard into our faces. It was blowing so hard that it kicked up one-foot waves coming towards us against the current. And then the skies opened, and it started to rain! Big fat raindrops pouring down.

The first crack of thunder seemed to be a ways away, but the second lightning bolt hit right in front of us, and the thunder followed almost instantly. Jeff and I had been paddling really close to the right shore already, but we both made a beeline to shore, and I immediately jumped out of my boat. I walked through the woods about 100 yards to meet up with where Jeff had pulled off. He had pulled off to the side but was still sitting in his kayak. I told him to get off the water. We sat in the pouring rain for 15 minutes or so until the lightning moved on. We were both already soaked to the skin, so I didn't bother getting out my raincoat.

We paddled another 20 minutes in a hard rain and into the wind, and spotted Bob, who had pulled over just 100 yards before our campsite. He'd stayed in his boat, floating in the river, and had never gotten out through the entire lightning storm.

We pulled up to the Suchens Campsite and unloaded our boats in the rain. It was a really crappy site consisting of a small clearing on a slope with barely enough room for our three small tents. The fire pit was hanging over a small bluff. The picnic table was sitting at an angle on a slope. Just a poorly designed site.

Since it was still drizzling and I didn't feel like setting up my tent in the rain, I decided to collect wet firewood and some birch bark. I was working on getting a fire started with wet wood and some damp cardboard, and Jeff came over to give me some fire building advice. He said he was known as a good fire starter. I hung a clothesline, put up my tent, and gave Jeff a crack at the fire. Twenty minutes later, Jeff had given up. I decided to give it one more try, and we've had a really nice fire going for the past two hours.

Because we'd had big Subway sandwiches for lunch at 3pm, no one was hungry for dinner, so I mixed up some powdered milk and had a bowl of granola. Jeff poured cups of Jameson Irish Whiskey for everyone, and we sat around the fire trying to dry out a bit. There was a nice rainbow when the rain finally stopped and an amazing sunset on the opposite side of the river.

We paddled just under 22 miles, crossed two portages, and bought the camping stuff we needed, so it was a good Day #8 on the river. Tomorrow our options are to stop at 23, 28, or 33 miles. Not sure if we can make 33 but that's what I'd like to try for.

4MPH All Day Long
August 4, 2019 - Day 9
Suchens Campsite to Jacobsen Campground – 33 miles
Total Mileage: 207.4

Great day!! 33 miles!! Furthest to date.

Going 33 today sets us up to end Tuesday in the town of Palisades, which has two restaurants and a county campground that might have showers!! We are all looking forward to that. The Jacobson Campground we are at tonight is run by the county and has eight sites. No one else is here. There's ice cold spring water coming out of a pipe, pit toilets, picnic tables, and really nice open campsites. The mosquitoes are bad. We've had weather reports of isolated thunderstorms and possible severe weather like yesterday afternoon, but it's almost 7pm, and so far, so good.

I cooked up red beans, rice, and wieners for Jeff and I for dinner tonight. Jeff had bought some pudding cups when we stopped with Sandy and Jeff a couple of days ago, so we each had one of those for dessert. Kind of a lame dinner, but it makes looking forward to a restaurant in Palisades that much more exciting.

This morning, when we got up, everything I'd hung on the line was still soaked and the tents were soaked too. The river was super foggy, and the temps cool. Suchens Camp was a sucky, tiny, sloping site, so we were all anxious to pack up and leave. It was super buggy too. I had oatmeal with craisins and granola chunks tossed in along with my Starbucks instant coffee while I packed up and was on the water by 7:15am. The other two followed about 30 minutes later.

I really enjoyed the first couple of hours in the cool fog. I used my ultra-lightweight canoe paddle, listened to *NPR*'s Sunday morning shows, and felt really relaxed. My shoulders barely ached at all today. And my lower back seems to be getting used to sitting all day too.

There were very few landmarks to watch for today, like eight or 10 miles between things, so some of the stretches seemed long. I got out to stretch my legs a couple of times but didn't see Jeff or Bob until I stopped at Swimming Bear Campsite 23 miles in, a little after 1pm. I sat on a sandy beach and had tuna and cream cheese on a tortilla, using the last of the cream cheese that I got at the hotel way back in Bemidji.

I only had to wait 15 minutes or so when Jeff came paddling around the corner with Bob right behind him.

The weather looked good, so we paddled another five miles to a boat landing. This was our next potential stop if the weather turned foul. The weather seemed to be holding, so Jeff and I paddled together for the next 10 miles and continued our conversation about various family members. Since we haven't really seen each other for probably 40 years, we have a lot to catch up on. Today we talked about his brother Duane and sister-in-law Ann, and a little more about his youngest sister Debbie. It's been really fun hanging out with Jeff and getting reacquainted.

The last 10 miles zipped past as we chatted away. We ended up averaging a little over 4mph all day which was awesome. There was a bit of a steady current all day, and virtually no wind in our faces. The sun was intensely hot in the afternoon, but I kept dumping river water on my head with my coffee mug, and hanging my hands in the cool water, and it helped.

I saw lots of wildlife today: a deer swimming across the river, an eagle, blue heron, some loons, and a little family of otters or minks.

Jeff told me this morning that Bob lost his spare canoe paddle a few days ago. He also lost his sunglasses yesterday. This is in addition to losing two phone charging cords and several less important items. So, Bob is going to need to pick up another paddle. Soon.

I also got a concerning text from Sandy and Jeff yesterday that said, "I am so glad you're paddling with Bob, and will be able to continue on with him. After observing him the last two days, I'm worried about him being out there by himself. "

So, there's that. . . .

Tomorrow we are planning on going 30 miles. There might be some thunderstorms tomorrow, so we'll keep our fingers crossed!

Nice People and More Rain

August 5, 2019 - Day 10
Jacobsen Campground to Sandy Lake Recreation Area Campground – 36 miles
Total Mileage: 243.4

Day #10 had some twists and turns that I didn't anticipate last night when we decided to paddle 31.4 miles today to the Libby Township River Campsite.

We're sitting here at a nice picnic table, at site #9 in the Sandy Lake Recreation Area campground. It's 8:40pm, and I'm trying to write in my journal when over walks Scott, an older guy from Duluth who said he likes meeting other campers. He just chatted away, with his Northern Minnesota accent, nice as can be. And then, over wanders Damon, the dad of the family camping two sites over, and he's carrying three paper plates, each with a few big, barbecued chicken thighs, and three ice cold cans of Grain Belt beer. I'd chatted with Damon when we arrived around 6pm. He also gave us some laundry detergent and agreed to take his clothes out of the only campground dryer so we could dry our clothes before the laundry room closed at 8pm. Super nice. Really friendly. Everyone sees our boats and mountains of gear and tents and are curious. That's how our day ended. Friendly campers, cold beers, barbecued chicken, clean clothes, and a hot shower.

How the day began was a little different! Up at 6am as usual. Pop-Tarts and a bowl of cold cereal with coffee. We all checked our various weather apps and got some conflicting reports. One said 30 percent chance of rain from 8 to10am, another said severe weather warning, and a third called for a chance of scattered thunderstorms, while the radar app I checked seemed to show that the storm would pass just north of us.

We heard a little thunder in the distance as we packed up, and it started to rain lightly. Based on the radar image I saw, I decided to take off at 7:20am.

After about 20 minutes of paddling in a light rain in my short-sleeved buttoned shirt and short pants, a bolt of lightning flashed in front of me, and I counted past 20 before I heard the boom. Too far away to worry about. Then a second one lit up the sky—10 seconds. And then a third flash followed almost immediately that I didn't bother to count. I bolted for the shore, jumped out of my boat, grabbed my lightweight raincoat and stern line, and climbed onto a steep shoreline that was covered with tall, wet grass.

I put on my raincoat and sat in the three-foot-high grass as it really started to pour. I texted Jeff and Bob that I was off the water, and then took a few selfies of my pathetic situation, texting them to Leslie and Seth hoping for some sympathy. When the lighting flashes and booming thunder got to four or five seconds apart, I scooted another few feet up the riverbank to get further away from the waterline.

And there I sat in the drenching downpour, waiting for the storm to pass. Bob and Jeff never even got their boats in the water and were sitting out the storm in the unlocked front porch of a little rental cabin at the campground. Smart.

The whole thing was kind of exhilarating. It definitely got my blood pumping and my mind running through scenarios of what I'd do if the storm didn't let up. We are each totally self-sufficient in terms of food, stove, fuel, tents, sleeping bags, etc., so, I knew that if I had to, I could hike into the woods, set up my tarp or tent, and put on some dry clothes—and even get a fire started. And God knows I have plenty of food!!

I texted Leslie and Seth and tried to generate a little sympathy. After about 30 minutes of sitting in the rain, it started to let up and seemed like the lightning storm was heading south. I waited another 10 minutes or so and then texted Jeff around 8:30am that I was getting back on the water. They ended up taking off around 9am.

It was a long paddle today, and I never really got into it until the last few miles. It could have been that I started my day soaked and sitting in a bunch of wet weeds for an hour, or it's more likely that I was lethargic because I'd spilled my entire cup of coffee into the bottom of my canoe right as I shoved off this morning. Big bummer.

I stopped at a boat haul-out spot around the 19-mile mark and made a spicy tuna and peanut butter tortilla. And, yes, it tasted as nasty as it sounds. I was just going for variety. I walked around a little and stretched my legs. The scenery today was mostly tall pine trees. Pretty, but monotonous. We only passed maybe a half-dozen houses over the entire 30 miles.

The campsite we'd picked as our stopping point was at 31.4 miles. At one point, I was pretty sure I had missed it, and then I was 100 percent sure. So, I turned around and tried to paddle upstream, and against what I thought would be a mild current. My futile upstream paddling only lasted a minute or two. I finally turned on my Aventa map phone app and realized that I hadn't quite gone far enough.

The campsite was completely overgrown. I tried to get out of my boat in two spots, and almost tipped over because the water right along the shore was too deep, and my feet didn't touch bottom when I was attempting to get out. Plus, the current kept pulling me downriver.

I checked the map and saw that the Sandy Lake Campground was about four and a half miles away. I checked online and saw that it had hot showers and a laundry. So, I waited about 40 minutes for Bob and Jeff to arrive and we decided to paddle on to Sandy Lake. It ended up being a great decision, but a long, long day. Instead of an overgrown, mosquito-infested site that would have been impossible to unload our boats at, we ended up at a great open site with lots of amenities and great people who stopped by to say hello.

Yesterday was the day that I started to feel in a groove. Settling into this trip and the rhythms of paddling all day, setting up camp, and getting up the next day to do it all over again. It has taken nine days. Today's 36-mile paddle was our longest mileage day yet. It was a lot of paddling, around nine hours, but beautiful, peaceful, with a little adrenaline thrown in to start the day. I like it better when things get a little challenging!

Brats, Beer, and Chips . . . Heaven

August 6, 2019 - Day 11

Sandy Lake Recreation Area Campground to Palisades County Park – 19.5 miles

Total Mileage: 262.9

Sitting by a nice, smokey campfire built in one of those campground fire grills. Keeping the mosquitoes away. I just finished cooking 10 brats over the fire. We each had three and one-third brats, shared an entire jar of pickles, and topped it all off with cold beer and barbecue potato chips. Great dinner!! The kind I dream about when I'm eating the typical dinners we've packed.

We're in Palisades, Minnesota, population . . . can't be much more than 200. Wait, I'll check. . . . Population: 167. We only had 19 and a half miles to paddle today. We all wanted to spend the night at the county campground on the river, located right on the outskirts of town.

Still, we were up at 6am. Jeff took a shower. Finally. I sort of lallygagged around, drank my coffee, and stared at the map for a while just to kill time. I could only stand the lallygagging for 15 or 20 minutes and then helped put all three boats in the water. I loaded up and took off around 7:40am. Bob left shortly after, and Jeff apparently didn't leave until almost 9am. I'm not sure why.

The first couple of hours were really relaxing.

I tried to focus on just relaxing as I paddled, observing the trees, enjoying the quiet. The sun was still low in the sky, and there was a bit of mist or fog just a foot or two above the river. I was consciously focusing on appreciating where I was. I'm paddling the Mississippi River! Day #11. It's so easy for me to get caught up in the mileage and just getting to the next destination—and getting the paddling over with. Trying to make the miles go by as quickly as possible. But this morning I focused on enjoying it.

I saw a couple of deer before they saw me. I passed several Canada geese who slowly floated off into the tall weeds, hoping I hadn't seen them. And then a big fat furry animal that scampered up a log. Maybe a woodchuck?

The scenery was mile after mile of trees. About two miles from Palisades there started to be some seasonal cottages and permanent homes on both sides of the river. And, finally, I arrived at the take-out point at the county park at around 12:45pm.

Bob was right behind me. We were completely unpacked, tents up, and organized before Jeff arrived. It's a really nice public campground with showers, picnic tables, and water. No one else is here. I scavenged a bunch of firewood, and we have a nice fire going this evening.

At about 2:30pm, we walked into town. It's just a three-block main street with a post office, town hall, two cafés, and two gas stations. We went to Gabby's Eats and Treats. The sign on the outside advertised meatballs. Jeff and I had talked about eating a light lunch and then coming back later tonight to the other café. Well, once Jeff heard the meatball special described, he was all in . . . potatoes, gravy, corn, and homemade meatballs. So much for the plan. I had a salad with grilled chicken strips, still with The Plan in mind.

After lunch, I stayed for two hours at Gabby's with the intention of doing some computer work, since the café had Wi-Fi. But I ended up talking with Leslie almost the entire time, which was much more fun. It was her first day of work at Unionville Elementary.

On the walk back to our tents, I stopped at the small grocery/gas station and bought brats, chips, and pickles to make for dinner. As I walked out the door a lady drove up to the store on her riding lawnmower. She had two empty grocery bags and had come to shop.

Only in a small rural town.

I did get some office work done tonight at the campsite. The brats turned out great.

Full and content.

There are storms predicted for tonight and early tomorrow morning. Our plan is to pack up and walk to breakfast around 6:30am. Then 31 miles!! That's a long paddle.

Okay, I can't see the lines on my journal paper anymore. Time to go to the bathroom, shave, brush my teeth, and head to bed.

Long Paddle and Whiskey. I Like It.

August 7, 2019 - Day 12
Palisades County Park to Aitkin City Campground – 31 miles
Total Mileage: 293.9

A 31-mile paddling day is a long day of paddling!

The first 10 miles went sort of quick. The middle 11 to 25 miles were really tough. And the last few miles went fairly quickly because they were the last few miles of the day.

I called Seth around mile 15 just so he could help psych me up. I put the phone on the floor of my canoe, and we talked while I paddled. It really helped! He's always so encouraging and supportive of my adventures.

Last night there was a massive storm. Around 1am. The wind started picking up, with gusts around 50mph (we heard later), that flattened the side of my tent into my face. And then the driving rain and lightning came. It poured for an hour and the lightning was non-stop . . . sometimes several different flashes in a second. It was wild! I worried a bit about the big tree branches above my tent. I wasn't in the mood to get smooshed. I thought that if it got much worse, I'd just head into the cinder block bathroom. It continued to howl with high winds for an hour and a half. Quite a storm.

In the morning, all was well. I kept thinking what it would have been like if that same storm had hit mid-day and we had to pull off into the weeds along the river and sit it out like I did the other day. That would've sucked.

I focused on the beauty today and tried to clear my mind as I paddled. Stroke, after stroke, after stroke. We passed a few old barns and farmer's fields, and not much else. At 12 and a half miles, I got out of my boat at a boat launch spot and walked around for a couple of minutes, but that was the only time I got out today. So, it was pretty much eight hours of straight paddling.

Around 1:30pm, I stopped and floated while I made a chub (summer sausage) and peanut butter tortilla, which was strangely excellent!

It was partly cloudy all day and only hit 80 degrees, so a nice day to paddle. But it was long and exhausting. Thirty-one miles is just a long-assed distance to paddle in one day.

The Aitkins Public Campground is nice. Big sites, nice bathroom, and shower. I got there at 3:25pm. Put my canoe on its trailer wheels and pulled it up to the closest site. I had my tent up and was all unpacked when Jeff and Bob showed up around 4pm.

We spent the entire evening at Block North Brew Pub. Awesome!! Had a great dinner of ribs and corn on the cob and good IPAs. After dinner, Jeff and I moved to the bar and had a couple of whiskeys while I did office work on my computer. I just said to Jeff, "Now this is the fun part of this trip!!" We got a lot of miles in and had a really fun evening. Can't ask for more than that!!

Paddling Rhythm

August 8, 2019 - Day 13
Aitkin City Campground to Bridge Tavern on Highway 6 (Crosby, Minnesota) – 25.7 miles
Total Mileage: 319.6

My best paddling day to date. I felt great all day. Felt strong all day long. I could have gone and gone and gone. Almost 26 miles and done by 2:45pm. It started out cool this morning. Low 50s. I had my insulated mug full of hot coffee sitting on the floor of my canoe.

I slept good last night. Who knows what it was, but I felt great all day. I had my AM/FM radio on for most of the day, listening to any station that would come in clearly.

The weather evolved from "Winds 5-10, gusting to 15" to "Winds 10-15" to "Winds 15," all coming out of the west. And today's 25 miles pretty much headed due west. So, a 15mph wind blowing against the river current makes waves, sometimes a foot high with white caps. Which required lots of extra effort, but I was strangely into it. The wind forced me to speed up my paddle cadence and really lean into each stroke. I think that wind and having to regularly change my paddle stroke kept me focused and interested.

Early on, two trumpeter swans flew overhead. And around midday a fawn started swimming across the river just as I was passing by, and the confused little guy turned in the middle of the river and started following me. He looked up, realized I wasn't his mother, and the made a beeline for the opposite shore. His mom stayed on her side, and I could hear her bellowing for her baby until I was out of earshot.

We paddled past lots more farmhouses and barns today, and scores of small river cottages. Way more than we've seen the past few days. So, there was lots to look at as I paddled along.

I spent a lot of time strategizing about whether to paddle straight into the wind or sneak along one side of the river or the other to stay close to the trees and out of the direct wind blasts. So, it was a fun and challenging morning moving back and forth across the river. It kept my mind occupied. I also brought a pad of paper and pen along in my shirt pocket today so I could take some journaling notes as I thought of them.

I felt really tuned into the river today. Just at peace. My paddling felt strong and fluid. I hope for more days like today. Tomorrow we have closer to 30 miles to go, so a bit longer of a day.

I did a little river math and figured that we'll hit Minneapolis on August 15th or August 16th. So, I called my mom to tell her I may be there for my birthday (15th) or my dad's (16th).

We passed lots more farmhouses and barns today and scores of river cottages. Way more than we've seen the last several days. So, there was a lot to look at as we paddled along.

The designated river campsite for this area is Lone Pine, but it was only 20 miles into our day, so we paddled another five and a half miles to a bridge where Highway 6 crosses the river heading to the town of Crosby. Bob called the Bridge Tavern, which is right by the bridge, and asked if there was a place to camp nearby. Whoever he talked with said we could camp right behind the bar.

I got to the take-out spot, a ramp just after the bridge, about 45 minutes ahead of Jeff and Bob. So, I strapped my boat onto my trailer wheels in shin-deep mud, and pulled it fully loaded up to the bridge so I could take some photos as they passed by.

Once they arrived, Jeff and I pulled our loaded boats about a half-mile down the road to Bridge Tavern and went straight inside for a cold beer before setting up our tents. Bob was a bit behind us. I'm done trying to help him get his boat on his trailer wheels. He has adopted a strategy of unpacking half of his gear before putting it on the wheels, pulling it out of the water, and then repacking. It takes forever, and neither of us was in the mood. And there was cold beer and an escape from the heat just ahead.

Jeff and I sat in the super friendly rural bar. Two of the regulars asked us lots of questions about our trip. Both riding big ol' Harleys. Probably Trump supporters, but super friendly. As we drank our ice-cold beers at the bar, we saw Bob pulling his boat towards the tavern, then right past the windows where we were sitting, and then on down the parking lot, with no sign of stopping. I jumped off my bar stool and ran to the door to yell for Bob, asking where the heck he was going. He replied that he hadn't seen our boats (which were sitting beside the tavern), so he thought we'd gone on further down the road. I still don't quite understand how he missed our orange- and gold-colored boats, sitting on trailer wheels, parked in a large mowed grassy area beside the tavern.

After setting up, we went back inside for dinner. I'm eating an amazing peanut butter and jelly hamburger that has jellied jalapeños. So good! I'm stuffed. This has really been the most fun part of the trip. Hanging out. Eating. Drinking. The paddling is just the means to this fun end.

Unfortunately, tomorrow morning will come too soon. We'll be off to Brainard. We have no place set for camping, but there is a city park right on the river that doesn't allow camping. However, a River Angel told Bob that she knows past paddlers have "stealth camped" at that park in the past.

I had a nice long call with Seth today. I tried to talk while I paddled, but it was too windy, so I pulled to the side and hung on to some weeds while we talked about the Kioti tractor that we want to buy this Fall.

I felt really into this river today. Really at peace. My paddling felt strong and fluid. I hope for more days like today.

I Am Iron Man[2] (or Iron Gut)

August 9, 2019 - Day 14
Bridge Tavern on Highway 6 to Brainard, Minnesota – 29 miles
Total Mileage: 348.6

Camping with a bar and restaurant in your front yard isn't a bad

---

[2] "I Am Iron Man" is a reference to a Black Sabbath song of the same name that was recorded in 1970. The song includes the line, "Has he lost his mind. . . ."

deal. Unfortunately, the Bridge Tavern doesn't open for breakfast until 10am. Jeff and I were up and at 'em just after 6am as usual. Oddly, Bob wasn't stirring. It was probably 20 to 25 minutes before he popped his head out of his tent. I was actually starting to worry a bit. He was just moving a little slower this morning.

We had to trailer our packed boats back over the bridge and down to the public landing area with a super muddy entry. Probably a foot-thick layer of fine silt enveloped my feet and wanted to keep my paddling shoes as souvenirs. I was on the water by 7:20am.

There was a four-foot-tall layer of surface fog on the water this morning that looked cool and kind of eerie. It added to the peaceful morning paddle. This is a really pretty section of river. Lots of big pine trees and a few small river cottages. I saw a few deer, heron, and several eagles before I stopped at a boat launch after around 12 and a half miles to stretch my legs. I had cold cereal and a Pop-Tart with my morning coffee, which doesn't seem to stick with me as well as two packets of oatmeal. So, when I stopped for my stretch, I made myself a peanut butter and chub (summer sausage) tortilla. I've been carrying this chub with me since I bought it in Deer Lake when River Angels Jeff and Sandy brought us to a grocery store 8 or 9 days ago.

Jeff has told me several times to throw it away, but I just opened it a day or two ago, and this morning I cut off a half-inch of the exposed end, and it tasted fine to me. No diarrhea yet.

As I was finishing my snack, Jeff paddled up, so we ended up paddling the remaining 18 miles together, which was fun. We haven't paddled together since early in the trip. We talked about his mom and her death, and a few other family related topics. I'd also been listening to *NPR* news earlier in the day, so as we paddled, we both launched into a Trump-rant about how much we hate this egotistical monster of a president.

As the miles passed, the wind out of the south picked up, and, you guessed it, we paddled pretty much in a southerly direction all day today. It only blew 5+mph but it was enough to make you have to work harder and paddle consistently so that you didn't drift backwards when you stopped for a break. After a couple of hours of stiff paddling, my right shoulder really started to ache.

At one point, we pulled into some weeds to take a break and have a snack. All day long it was sunny, and the breeze felt good. Just a pain to paddle into. The final 18 miles or so we started seeing more and more fishing boats and speed boats on the water. Even a few jet skis. As we neared Brainerd, the shores were lined with nicer and nicer homes. A few of the boaters cut their throttle when they passed us, so they didn't cause a big wake, but most just blasted on past, waving to us as their two- or three-foot waves crashed into our tiny boats. Each time, I tried to swing my canoe around and steer into the oncoming waves so they wouldn't tip me over or fill my open boat. Jeff's kayak, of course, just cruised right through.

Usually around mid-day, when I'd stop for a break, Bob would cruise up behind us, never too far back. But not today. I stopped several times in big open stretches of water to look back for him, but I never saw him.

At around mile 26, we got to the Brainard Dam. Another one built to power a massive paper mill. A 210-yard easy portage, followed by a pleasant and quick half-mile paddle to Kiwanis Park.

Our plan was to paddle to Kiwanis Park in Brainerd and see if there was a hidden spot to camp. We knew that camping wasn't allowed but thought that if we could find a flat, forested spot, we'd just hang out until dark and then set up our tents.

Jeff and I pulled into the park around 4pm, and it was just too populated. Lots of cars in the parking lot, a playground, and no obvious place to hide three tents. So, we quickly decided to find the nearest hotel, and then figure out what to do with our boats. I got two rooms at the Quality Inn, a mile and a half away. Jeff called a River Angel contact, Vicky, to see if she could help transport our gear and boats to the hotel. She said she'd talk with her husband and call us back. After 45 minutes, I got antsy, so I called a Lyft driver, and he took me and most of our stuff to the hotel. About the time I got to the hotel, Jeff texted that Vicky's husband Jon was on his way.

He made two trips to haul our boats. Bob showed up around an hour after we'd arrived. Jon was super nice, but he clearly was doing this to help out his wife Vicky. She's the River Angel who wants to help paddlers, and who occasionally wrangles her husband in to help.

The Quality Inn only had two-room suites, so Jeff got the room with the bed, and I got the room with the fold-out couch. Within twenty minutes, tents, flies, sleeping bags, air mattresses, etc. were spread out all over the two rooms, and the place looked trashed. I did a load of laundry and washed my sleeping bag. And took a shower. Jeff ordered pizza and had it delivered to the hotel and walked to a local gas station to buy some beer to go with the pizza.

Mid-pizza, Bob came into our room and sat on the couch and announced that he was going to take a half-day or full day off tomorrow. Jeff and I were shocked!! Bob looked really tired and worn out and said that he had several errands to run in the morning and then wanted to take Jon and Vicky out to lunch. By the time he was done talking, he'd decided to just take the whole day off and then spend the night at Jon and Vicky's. We were stunned.

It all made sense. The last few days, Bob has been dragging. Slow to get up in the morning. I think the pace was just a little too much for him and he finally just needed to take a day off. He's 77 years old for God's sake. Good for him. I'd totally convinced myself that Bob would be with us for the duration.

During the evening, my brother Aaron emailed a 15-minute video about the Mississippi River that aired on CBS this past July. Really interesting. And Seth sent several photos from the cockpit of his commercial jet, of the river near St. Louis, that showed dozens and dozens of barges backed up for miles due to the heavy flooding that was taking place downriver.

I tried just sleeping on the couch, but it was too uncomfortable, so around midnight, I pulled out the hide-a-bed. It was worse!! And the only blanket (no sheets) was coarse and nasty. Oh well!

Where's Bob?

August 10, 2019 - Day 15
Kiwanis Park, Brainard, Minnesota, to Fletcher Creek Campground – 28.5 miles
Total Mileage: 377.1

I'm pooped!

It was a beautiful day of paddling. Some huge expensive river homes for the first few miles out of Brainerd, and then miles and miles of forest. There were several small bass fishing boats on the water and four rowing dories that looked like guided boats with a fit rower sitting in the middle of the boat, and then an old fat guy in the bow and another in the stern with fishing poles and beers in hand.

We paddled through a couple of small rapids and the flow really sped up over the last few miles. So, that made the water a little squirrelly and made the paddling a lot more fun and interesting.

No Bob today.

Feels weird.

For much of the day we passed the Ripley Army Base on the right shore. There were military exercises going on and I heard machine gun fire throughout the day.

We got a few light sprinkles that lasted 30 minutes or so, and there was a slight breeze coming out of the south and blowing into our faces that slowed things down a bit. But, overall, it was a great paddling day. I saw several eagles, lots of deer, and got some great close-up photos of a blue heron.

We found the designated river campsite we'd been heading for: Fletcher Creek. It was a cleared site with a picnic table, and looked well-used, but required landing our boats along a steep seven-foot-tall, eroded bank. We had to get out into waist deep water and toss our shit up the bank onto flatter ground.

Pictures can only tell what a thousand words cannot, but as Jeff was beginning to throw his countless dry bags out of his kayak and up over the bank, he lost his balance and fell backwards, windmilling his arms, falling into water that was over his head. It was pretty hilarious.

Unfortunately, the site was super mosquito-ey!

We set up camp, doused ourselves in bug spray, and it wasn't even 4pm. The thought of spending the next five hours sitting at a buggy campsite wasn't very appealing. So, I had a slight cell signal and scoured a Google map of the surrounding area for some place to hang out. I found the Boomerang Bar just over two miles away. I talked Jeff into packing up our charge cords, phones, and electronics, and walking there for dinner and a beer. Definitely a way better option than getting chewed up all afternoon and evening.

We had to follow a trail through the woods that led us to a small parking area by the river. Google maps took over from there. Once we got to a road that had a little traffic, I stuck out my thumb, and we ended up getting a ride the last half of the way.

So, here we are. I'm journaling. Jeff is blogging. I had a Reuben sandwich and bowl of wild rice soup. And an IPA of course. Way better than sitting in camp getting chewed up. We'll walk back in a bit, and I'll probably go right to bed. It's already 6pm!!

Where's Bob?

Sauk Rapid Angels

August 11, 2019 - Day 16
Fletcher Creek Campground to Two Rivers Campground – 25.5 miles
Total Mileage: 402.6

What an amazing day, only because of the people we met along the way.

I was on the water by 7:15am after a stunning breakfast of oatmeal and a sweet roll from the Quality Inn that I'd been carrying with me for far too long. I tipped over my full cup of instant Starbucks coffee while packing up my boat (again) and pushed off the shore with NO COFFEE! Crap.

Rather than load his boat by hauling his 25 dry bags down the steep seven-foot drop-off that separated our campsite from our boats (and after Jeff fell in over his head yesterday trying to unload his boat when we got here), I helped Jeff haul his boat out of the water, up the bank and into the campsite so that he could load it on dry land. He then wheeled his full boat down a 200-yard trail to the public boat landing.

Well, apparently that didn't go so well. He texted me that it took him almost an hour after I'd left to get his boat hauled and loaded. I'm not sure what all the problems were that he encountered.

I paddled the eight miles into Little Falls and got to the portage that led around the dam by 9:30am. I unpacked my boat, carried it up several steep steps to the sidewalk, went back several times for my gear, and then put my canoe on the trailer, reloaded, and pulled it about 300 yards towards the put-in spot that was just below the dam.

About halfway there, I looked over and saw three old guys sitting on the small patio of a unit in what looked like a senior apartment building. One of the guys yelled over, "Hey, do you want a cup of coffee?" I left my loaded boat, and walked over to meet Keith, Daryl, and Leland, who were spending their Sunday morning people-watching. My kind of guys.

Knowing that Jeff was on hour behind me, I stood and had two cups of horrible weak coffee poured by Daryl, who was so old and shaky that his hands could barely pour the coffee. I think this must've been his patio. We all chatted away for the hour. Super fun.

Jeff eventually texted that he had arrived at the portage, so he strapped on his portage wheels in the water, and I helped him lift his boat up onto the sidewalk. I suggested to Jeff that he introduce himself to the guys while I walked to a gas station and bought some Mountain Dews and two cups of ice. By the time I got back, Jeff was sitting there like he was one of the gang.

We eventually reloaded our boats while Leland and Keith watched. And off we went. Daryl couldn't make the trip.

At mile 16 and a half, we got to Blanchard Dam. Our maps said that it was a 600-yard portage, but a river outfitter that I met in Little Falls who happened by while I was talking with the three old guys said it was more like a mile and a quarter! With several steep ups and downs (600 yards is only one third of a mile).

It turned out to be more like three quarters of a mile, but there were several steep hills and Jeff and I had to help each other to push/pull our partially loaded boats along the rocky dirt trail. We initially tried to pull our boats mostly loaded to save on trips, but the trail was so steep and rocky in places that we ended up unloading piles of gear along the way that we had to go back for.

And it was hot! Mid-80s at least. And sunny.

I felt like I was cooking in the heat. The length of the portage, and having to make three complete trips, really took it out of me. It took almost an hour and a half to haul everything across. By then it was 3pm, and we still had nine miles to go.

I made a peanut butter and cheese tortilla with the cheese I've been lugging around for almost 10 days. The cheese was super mushy in the heat, but was still unopened, so how bad could it be?

Last night I started getting nauseous around 11pm and felt like I was going to throw up for about an hour. I even got the sweats. It might have been the 10-day-old summer sausage that I had for lunch earlier yesterday.

I bought the cheese at the same time as the summer sausage, so am fully expecting to get sick again tonight.

The last nine miles were really hot and sunny, and after the long portage ordeal, we were both pooped and felt like we'd had too much sun. So, the final nine miles seemed to take forever. My shoulders were really aching, and it took a while to get back into a rhythm after the long portage break.

Since we started the trip two weeks ago, Jeff had been in contact with a River Angel named Lee, who lives in Sauk Rapids, just north of St. Cloud. We originally thought we might paddle all the way to his house today, 41 miles, but after the hour chatting with the three old guys in Little Falls, two portages, and the intense sun, we scaled back our plans to go 25 and a half miles to a private campground in Twin Rivers. Lee agreed to pick us up there.

Two Rivers Campground and Tubing was a beautiful campground with a nice sandy beach on the river where we pulled out our boats, incredible well-spaced campsites, plus mini golf, a big swimming pool, a rec hall, lodge, etc.

There were signs all over the beach and campground saying, "No Trespassing" and "For Guests Only" so we walked up to the "lodge" and asked if they'd let Lee drive through their "Guests Only" gate to pick us up down by the river where we'd pulled out. Their answer was, "Sure, if he pays the $24 entrance fee." Really?

So, we paid it rather than haul our loaded boats uphill a quarter mile from the river to their entrance.

Lee is a great guy. He and his wife Patty started helping out paddlers last year, and already have plenty of stories to tell about paddlers they've met. Their home was completed just last year, and is beautiful, with marble counter tops throughout and big windows facing the river. Just amazing.

They invited us to stay in their guestroom. And Patty, who knew in advance that we were coming, prepared grilled steaks, baked potatoes, a bean hot dish, fresh baked crescent rolls, and on and on. This was served after the huge fruit tray, veggie tray, and meat and cheese trays that she'd prepared for hors d' oeuvres. Their neighbor John even came over to meet us.

And the best part? Lee has an entire refrigerator FILLED with dozens of obscure IPAs. It was the most exotic array of IPAs I've ever seen. FILLED. Hundreds of cans and bottles. It was like an IPA museum.

We ate and chatted through the evening. Patty started serving delicious after-dinner gin and tonics. And we were up until 11pm. What amazing people. I was worried, based on my own stereotypes, that they were republicans, but right towards the end of the evening, Patty said something about hating Trump, and found that I loved them even more!

Patty even did a load of our laundry.

Just super people who let total strangers into their home out of the goodness of their hearts. Lee suggested not only taking us back to where he picked us up tomorrow morning, but he agreed to keep our bags in his truck, and then meet us at our second portage tomorrow so we can paddle empty boats and have super quick portages.

Great day!

What Will the Neighbors Say?

August 12, 2019 - Day 17
Two Rivers Campground to Clearwater, Minnesota – 29 miles
Total Mileage: 431.1

It's already 10pm and I need to get to sleep. I never seem to have enough time in the evening to write, read, and do some office work. I really need to buckle down on my work. Things are starting to pile up.

We're camped on a little patch of grass next to Clearwater Outfitting in Clearwater, Minnesota, right on the river. Jeff had called the owner of the store in advance to see if we could camp here. And then we met him when we arrived just before the store closed at 6pm. The owner said, "Whatever you do after we close at 6pm, we don't know about." He also asked that we stay out of the view of the neighbors, since their business isn't zoned for "overnight lodging." (Apparently, we aren't the first paddlers to ask to spend the night.)

So, we waited until dusk to set up our tents, and we'll have everything taken down and packed by 7am.

Otherwise, here is a synopsis of our day:

- Lee cooked a nice breakfast of eggs and bacon and dropped us back off at our take-out point. He kept most of our gear in his truck.
- We paddled 12 miles to Sartel, did the half-mile portage around the dam, and then paddled another seven miles right through St. Cloud to the big dam.
- Lee met us there and drove us to the other side of the dam. He also took us to a Quick Mart to buy some crappy, deep-fried lunch food (not Lee's fault), and we ate our lunch with him alongside the river before we loaded up our boats and took off.
- Back on the water at 3:45pm for our final nine miles to Clearwater. We passed a great campsite called "Boy Scout," so that we could get to the canoe/kayak store to camp. Not sure why?
- I called ahead and bought two tie-down straps and a Kevlar skid plate repair kit to patch the gouges in my bow and stern from inadvertently smashing too many rocks.
- Going through Sauk Rapids was a little scary in my open boat. I smashed a couple of rocks broadside and almost tipped over. It was the fastest rapids we've paddled through yet.
- I texted Bob to check on him and heard he was back on the river yesterday, and then again today.

Jeff and I spent a lot of time tonight looking ahead to plan out the next few days of getting to and through the Twin Cities . . . take-out points, transportation logistics, etc. Jeff's wife Chris will be there, and my mom and dad live in St. Paul, so we want to plan ahead when and where we'll be picked up. We may get to St. Paul as early as Wednesday evening (in two days). I need to figure out how to get my boat picked up, hauled, and returned the next day, so that I can spend the night at my mom and dad's.

We paddled 29 miles today, which is great given that we weren't in the water until 9am and had two portages.

Not So Nice People and More Rain

August 13, 2019 - Day 18
Clearwater, Minnesota, to Elk River, Minnesota (in the weeds) – 32 miles
Total Mileage: 463.1

It doesn't seem like there is much to write about today. We're sitting on our collapsible camp chairs along the river, two miles south of Elk Rapids. There was no place to camp in town and no nearby hotels in Elk Rapids, which we knew when we started the day. We'd originally hoped to stop at a riverside park in town, and then hang out at a picnic table until dark to set up our tents. But when we actually got to the park, it was only 3pm, and the park was right next to a very busy road with lots of people enjoying the park.

I called the two nearest hotels, both of which would have been a hassle to lug our boats and gear to, and both were full. So, we paddled on.

Shortly after we paddled away from the park, big, dark storm clouds started closing in around us, with a forecast of rain for two or three hours along with lightning. So, we started to get a little anxious about finding a place to camp and getting our tents set up before the skies opened up. We heard some thunder creeping up behind us and started earnestly scanning both sides of the river for a small clearing and place to haul our boats out.

We paddled past a trailer park and saw a lady outside bringing in her lawn chairs before the storm hit. "You're going to get wet," she yelled. (Funny.) I asked if there was some place we could set up our tents? "No" she replied, "This is a private trailer park." (Not super helpful.) I said, "We're trying to get off the river before the lightning gets here. Is the owner of the park nearby, so we can ask them?" "I'm an owner," she replied. "But you can't set up here."

Fuck it. We paddled on.

About two miles out of Elk River, I spotted a small break in the trees and a tall grassy area on the right bank, just as the lightning started to strike. We quickly pulled off, threw everything up onto the shore, pulled our boats out of the water, and just as I got my tent set up in record time, it started to rain, hard. I tossed my sleeping bag, sleeping pad, electronics, and books inside, and closed the dry bags that would stay outside, and hopped in the tent at 4pm. And it proceeded to pour, and thunder, and lightning for the next two hours.

It's such a nice feeling to be in a tent, and out of the elements. It was warm and muggy, but way better than being outside. I made a tuna and tortilla roll-up, checked my email, texted Seth, Tyler, and Leslie, read my book for a while, and even dozed off.

Right now, Jeff and I are sitting outside on our small camp chairs. Its 7:15pm. Looks like more rain might be coming soon. But, at the moment, it's nice outside. And very few bugs!! I made a rice and chicken with sesame ginger teriyaki sauce for dinner. It was pretty good. Now I'm drinking hot cocoa, coffee, and powdered milk all mixed together. I was in the mood for something sweet.

This morning we started packing up early so that we could get our tents down before the neighbors saw us. Around 6:30am, just as we got our tents down, it started to rain, so we moved our breakfast and gear underneath a big tarp they had set up for events, where we nibbled at breakfast while we finished packing up. Unfortunately, I accidentally left my awesome insulated coffee mug there!

We were paddling by 8am. I waited for Jeff to get all packed up so we could paddle together today. We passed lots of nice summer homes and cottages, but still had long multi-mile stretches of trees.

It was sunny, but only in the low 70s. While we paddled, I worked on some pick-up and drop-off logistics by phone and text with my mom and my niece Emma. I think it'll be exciting to paddle through the Twin Cities over the next two days. We should be on the north side of Minneapolis, at the St. Anthony Falls lock and dam, by late afternoon tomorrow. I've contacted a marina in south St. Paul that said we can tie off our boats there on Thursday night. So, that's all taken care of.

At around mile 16 today, we hit the town of Monticello. We stopped at a park and ate and stretched our legs a bit. We also walked to a Walgreen's and bought coffee mugs since Jeff lost his yesterday, and I left mine behind this morning. I'm also looking for a new portable AM/FM radio. My cheapy radio stopped getting FM stations for some reason. And AM is pretty much just Rush Limbaugh or Christian preachers.

No radio at Walgreen's, so I'll check Target in the Twin Cities.

We paddled another 14 miles to Elk River, and then the final two miles to our campsite in the weeds.

The other event worth reporting started when Jeff realized while packing up this morning that he'd lost his portage wheels. We reconstructed yesterday's events and surmised that when we lifted Jeff's boat off the portage wheels and into the back of Lee's trailer at the second portage yesterday, we must've just left the wheels sitting there in the parking lot.

Jeff called the Botanical Gardens where we'd met Lee, and sure enough. . . .

Jeff then called Lee to see if he was willing to go to the Botanical Gardens, pick up the wheels, and then meet us at the park in Monticello.

About two hours later, Lee called from the campground pick-up spot where we'd originally met him two days ago to tell Jeff that his wheels weren't there. Jeff swears that he told Lee the Botanical Gardens, but Lee drove 30 miles in the opposite direction to the wrong spot. Uh oh.

Jeff proceeded to ask Lee if he could head to the gardens and grab the wheels, and Lee responded, "No. This isn't what I signed up for."

Double uh oh.

Jeff, of course, graciously said, "Thanks anyway."

So, apparently, we reached the limits of Lee's River Angel benevolence.

Another storm is coming our way. It's 7:40pm. We'll definitely get enough rest tonight. Tomorrow—Minneapolis!

Home Sweet Home
August 14, 2019 - Day 19
Elk River Camp to Boom Island Park in St. Anthony Falls, Minnesota – 28 miles
Total Mileage: 491.1

Made it to Minneapolis!!!

It was cool to paddle around a corner and suddenly the entire city skyline came into view. And then it was industrial cement plants, a huge metal recycling plant, rusty steel railroad bridges, and the low constant hum of city traffic.

Sounds of the city.

Knowing that my mom, dad, and Emma would be coming to Boom Island Park to meet me, and then we'd be heading to my mom and dad's for dinner and to spend the night, added some extra motivation to my tired, sore shoulders and neck.

After 19 days, my shoulders and neck still start out really stiff and sore each morning. I can usually paddle them into not being so sore after an hour or so, sometimes with the help of some Advil. And then they start getting sore again at some point later in the day (probably after the Advil wears off).

When we got up this morning, everything was soaked. It must've rained quite a bit last night. I was wide awake at 5:30am, out of my tent at 6am, and paddling by 7am. I didn't feel like sitting around in the wetness and was anxious to get to the Twin Cities. My tent fly was soaked, but I just stuffed it in the stuff sack.

Crumbled Pop-Tarts and coffee for breakfast.

At mile nine, I pulled off at a park to poop. There was a group of twenty or so getting ready for a day-paddle that was being sponsored by Midwest Mountaineering and Wilderness Inquiry. I'm sure that they had no idea that I was paddling the river and on Day #19.

Jeff and I met up at the Coon Rapids Dam. I looked around and couldn't find a portage sign, so I picked a spot that looked kind of worn down. Then I walked the 600 yards to the other side to make sure it connected to the downriver side of the dam. When I got back, Jeff was pulling in. There was a park and small beach nearby, and lots of joggers and cyclists and dam visitors milling around.

We stopped for a snack after hauling our boats and gear across the portage. I made a peanut butter and beef jerky roll-up with my last remaining tortilla.

Jeff had to use my portage wheels since his were still sitting at the Botanical Gardens.

After the dam, there were two sets of rapids noted on the map. Since some of the rapids we hit yesterday and the day before weren't marked on the map, I was concerned that these marked ones could be big. So, I unfurled my spray deck for the first time on this trip and fastened it down to cover the entire boat.

Coon Rapids came and went without as much as a ripple, and the second marked set of rapids a couple of miles downriver were a big nothing burger.

A few more miles of paddling and the Minneapolis skyline came into view. It was pretty amazing to round a corner and have the entire city seemingly pop up out of nowhere. We paddled past several industrial sites and were pulling into the Boom Island boat launch by 2:30pm, about two hours ahead of when I'd asked Emma to pick me up.

Chris showed up right as I was pulling my boat out of the water. After we unloaded our boats, Jeff and Chris went to a local brewery and brought some beer back. So, we sat at the park, drank good beer, and waited for Emma and my parents to arrive.

Back at my mom and dad's apartment, I unpacked all of my bags in their spare room. It looked like a bomb had exploded. A complete disaster. Stuff everywhere.

I threw my soaking tent and fly into the dryer and did a load of laundry. We parked their car, with my canoe tied on top, in the parking garage for safe keeping.

After a nice quiet dinner of spareribs and baked potatoes, followed by a shower, I was sound asleep by 9pm.

Happy Birthday

August 15, 2019 - Day 20
Boom Island Park to River Heights Marina – 22.2 miles
Total Mileage: 513.3

It's my birthday. Fifty-eight years old.

Jeff and I decided to get up early and get our paddling done for the day, so that we could enjoy at least part of the afternoon with our families.

Nephew Nathan met me at my mom and dad's apartment at 7am and we drove the car with my boat and just a day bag to a put-in spot just downriver from St. Anthony Falls at the University of Minnesota boating center. Jeff and Nick showed up shortly, and we were on the water by 8:30am and paddled 22 miles by 3:30pm.

Seven hours on the water. Shorter than usual, which was nice!

Twenty-two miles is still a long way to paddle. The scenery was much prettier than I expected. I thought the 10 miles or so between Minneapolis and St. Paul would be industrial, but it was actually the opposite. All forest and a couple of parks. No sign of city buildings at all. A few people walking and biking. Some guys fishing along the shore. It was really a pretty and peaceful paddle.

About halfway between Minneapolis and St. Paul, we reached Mississippi River Lock and Dam #1. Our first river lock. River Mile 847.9 according to our Army Corps map. The first of 29 locks that we'll pass through on the Mississippi.

Both Jeff and I have read about people paddling through these massive lock and dam systems in tiny boats. Locks were built by the Army Corps of Engineers so the big barges pushed by tugboats can navigate the changing water level of the river and move commerce up and down the river from New Orleans to the Twin Cities, and to other major connecting river arteries, like the Ohio, Tennessee, and others.

We'd read about an option suggested by some paddlers to call the lockmaster 30 minutes or so before getting to the lock to let them know you were coming and to inquire about boat traffic and how long you may have to wait for your turn. We'd also read about the pull-cord that is supposedly attached somewhere along the outer lock wall, just before reaching the huge gates. We'd read some Mississippi paddling authors who wrote about how intimidating and dangerous the locks can be in a small boat, and others who wrote that they'd found a way to portage around every single lock on the river. And we met paddlers, like at the boat store in Clearwater, who said they had no interest in paddling their tiny kayaks or canoes through any of the locks.

So, as we approached Lock #1 just south of Minneapolis, I definitely felt some trepidation. But thinking back on it now, all I can remember is how much fun it was!! Pulling the cord and hearing the booming fog horn, listening for the lockmaster's voice to tell us how long we'd be waiting, watching the massive doors slowly open, paddling into the lock and grabbing hold of one of the many ropes hanging down for small boaters to hang on to, so you don't accidentally drift towards the front or rear doors, and then feeling and seeing the water level slowly, almost imperceptibly drop, as the concrete wall on each side seemed to grow higher and higher. And then, after the big doors open downriver, we paddled out into a new river world.

Super cool. Super safe. Relaxing. Exciting. Fun.

Can't wait for more!

The view of the city skyline heading into St. Paul was really cool. And then paddling through the city and past the big paddle wheel boats was fun. The *Jonathan Paddleford* was just pulling away from the dock with a boatload of passengers as we paddled by.

It was fun to see familiar sights of downtown St. Paul. Things I've seen from the road, but never from the river, over the past 40 years—paddle wheel boats, the sand bluffs, Mounds Park. This was also our first experience paddling alongside loaded and empty barges and tugboats. I felt nervous as we got close to the barges, especially a few that were being moved around by the tugs. Which way are they going? Are they turning? Can they see us?

The river heads north as it passes St. Paul, and when it turned east and then south, we were hit with a stiff breeze that blew into our faces for the remaining 10 miles. That sucked. It took the energy and motivation right out of me! I was really struggling the last few miles.

We finally pulled into River Heights Marina after 22 miles and asked if we could leave our boats for the night. The waves were starting to pile up from the wind, and we were getting blown backwards whenever we stopped paddling. So, it was great to finally stop for the day. I'd originally hoped we could go another five miles or so to another marina that I'd contacted yesterday, but what's the point?

We took an Uber to my parent's place and then drove to Cabella's and Target to resupply. I bought a new short-sleeve button shirt to paddle in, and a t-shirt to sleep in. And lots of food to replace my breakfast and lunch stock that I've grown totally sick of after 20 days. My food bag is now so heavy!! No more buying food. I have enough for a month.

Mom, Dad, Jeff, Chris, Nick, and I went out for a birthday dinner at Olive Garden for me (today) and my dad (tomorrow), and then home for birthday cake and ice cream. Mom got me a big Green Bay Packers pillow. Awesome! But too big to bring with me in the canoe. It turned out perfectly that we were paddling through St. Paul on my birthday.

Beach Fire Bliss

August 16, 2019 - Day 21
River Heights Marina to River Mile 807 – 23.5 miles
Total Mileage: 536.8

We decided to start a little later today. We met at the marina at 10am and pushed off at 10:40am.

I had a hard time today.

My right triceps really ached. I ate a total of eight Advil throughout the day. My boat felt heavy and slow. Too much new food?

We locked through Lock #2 today, which was fun. Again, we were the only boats in the lock. We had to wait about 15 minutes when we first arrived and pulled the "we're here" cord. The water was really choppy and squirrelly on the upriver side of the lock, so it was hard to stay in place while we waited for the huge doors to swing open.

This lock only dropped about eight feet, so it didn't take very long for the water level to drop before we were paddling out, and then past the town of Hastings, Minnesota. We pulled off at a crappy little marina to buy cold pop and stop for some lunch.

We paddled another nine miles before pulling off to camp. A big barge passed us heading upriver, and dozens of speed boats zipped by, throwing up big two- to three-foot-tall waves. Each time they zip past, we have to turn our bows into the oncoming waves and ride them out before turning and heading back downriver. Sort of obnoxious.

The scenery is beautiful. The Mississippi is now a big wide river. Tugs pushing barges up and down the river. Not as many today as yesterday.

We passed some beautiful homes sitting high on the bluffs.

In Hastings, the St. Croix River joins the Mississippi, and now Wisconsin is on the left bank as we head downstream. Up until now, for the first 520 miles of the river, we've been solely in the state of Minnesota.

We are camped just up into the trees next to a huge sandy beach. It's the nicest site we've had so far. Jeff started a nice beach campfire and served a baguette with Havarti cheese and salami for dinner along with some Surely Brewing Company beer. It was a really relaxing evening, sitting in our camp chairs on the beach by the fire.

A huge tugboat pushing 12 barges just pushed upriver. Pretty cool.

I hope I'm a little more into it tomorrow. I think being in the Twin Cities for two nights threw me out of my rhythm.

The river has changed so much over the last few days. It's now a wide working river, and our tiny little boats seem so insignificant here.

This beach is a beautiful, serene spot.

Still no other canoes or kayaks on the river.

Just us.

Day 22 in Two Completely Different Episodes

August 17, 2019 - Day 22
River Mile 807 to Star Point Campsite – 30 miles
Total Mileage: 566.8

Episode One: Five Rules to Lock By

Today was a tale of two days wrapped into one. The first half was glorious. The second half sucked. Bad.

Everything was covered in fine sand when we packed up this morning. But it was a great campsite and a perfect morning. A hot cup of coffee. Sitting in my camp chair. Calm river. Very little breeze.

Lots of trains passed on the Wisconsin side of the river during the night, but otherwise I got a good night's sleep.

It was a 10-mile paddle to Lock #3. I left before Jeff, so got to the portage before he caught up. I got out of my boat and walked around to stretch my legs. There were lots of eagles this morning. I found a small eagle feather floating in the river yesterday, and I found a second eagle feather while I was stretching.

Once I saw Jeff in the distance, I called the lockmaster to see if there were any big barges coming up from below the lock. We were told at Lock #2 that these big barges can take an hour and a half to two hours to lock through. The lockmaster said to just paddle up and wait for the green light. There was no boat traffic heading in either direction, so he locked us right through. So far, early on this Saturday morning, there hadn't been any yacht or fancy boat traffic. It was a very peaceful and enjoyable paddle.

The guy at the lock was great. As he stood on the side of the lock and dropped ropes down to us to hold onto, he said, "I have five rules that you need to adhere to when you are in my lock: One, no wakes. Two, if you have chalk to write something on the concrete lock wall while you wait, no body parts, and nothing pornographic. Three, play nice in the sand box. Four, don't damage my nice concrete walls with your boats. And five, wait for the horn before you exit."

After exiting the lock, we paddled another six miles to the town of Red Wing, home of the famous shoe factory. The boat traffic picked up, and there was an increasing parade of speed boats and giant yachts racing past us. Not one slowed down, and all of them threw up two- to three-foot wakes that we had to turn and paddle directly into to keep from tipping over. Really annoying.

We wanted to stop in Red Wing for lunch. We were hoping at least for a picnic table to make our tortillas and get out of the boats for a bit. Right on the south edge of town we spotted a marina, so we decided to pull over. There were two high school aged young ladies standing out on the main float who waved us in and asked if we wanted to buy some cold water or pop.

When they learned that we'd started almost 600 miles back, and 22 days ago, they called their boss Cindy down to meet us. Cindy has been managing the Red Wing Marina for over 20 years and was super nice. She immediately got us each a free cold beer and offered us a place to sit on big metal chairs underneath a patio umbrella.

Jeff ordered a pizza, and what started out as a plan to eat a couple of tortillas at a picnic table quickly became pizza and beer with free beer refills at a marina. Cindy and several of her high school summer staff asked all kinds of questions as we ate. They offered to charge our phones and even offered a shower (which I now wish I had taken advantage of). It was a great stop. Cindy wanted pictures of us, and I got a good one of her and Jeff.

The sun was out, and the temperature was climbing into the mid-80s . . . hot for being on the water. The heat and the beer made it hard to find the motivation to get back in our boats and paddle.

Episode Two: Getting Swamped on Lake Pepin

After five more miles of paddling into the wind and putting up with speedboat wakes that made every effort to swamp us, we hit the north end of 20-mile-long Lake Pepin. The parade of speed boats and yachts heading both directions kept picking up on this sunny summer Saturday. And the wind was picking up as well, causing its own waves to contend with.

After a couple of miles of paddling hard on Lake Pepin, I told Jeff I was going to pull over for a short break to stretch my legs. So, Jeff paddled on ahead. It was about 3pm. I figured we'd get to the 30-mile mark, where Cindy suggested we stop, by about 5:30pm. A 10-hour paddling day.

Back in the boat, and with Jeff far ahead of me and out of sight, I spotted a shallow section up ahead, and the waves were really piling up into white caps. It looked really shallow and turbulent, so I headed left, out into the main channel to go around the shallow section. And that's when my day took a U-turn.

Just as I got beyond the shallow section, a huge yacht came blasting past, only 20 yards away. I didn't have time to turn my bow into the oncoming wake, so a four-foot-tall wave came crashing over the side of my canoe, instantly filling it half-way with water. The first wave was immediately followed by a second large wave that completely filled my boat.

I was swamped!!

Anyone that has ever tried to paddle a canoe that is half-way or more full of water knows how tippy it becomes. It is almost impossible to keep it from rolling over.

So, here I was, a good half-mile from shore, still in the main boating channel, speed boats zipping all around, boat full of water, scared that my computer bag, and everything else was getting soaked, and it was all I could do to keep the boat from completely tipping over and dumping everything into the lake. And I didn't want to be floating in the water trying to gather my gear and swimming my boat the half-mile to shore while speed boats full of half-drunk passengers whipped past, not paying attention, and possibly running into me.

I was too scared to be angry.

I tried to bail out water with my pee bottle, but boats just kept screaming past me, so I had to concentrate on keeping my canoe from tipping over. I bailed out 15 to 20 water bottles full of water, and finally decided that I needed to paddle all-out for shore, while stopping to stabilize my canoe every time another boat wake pounded me.

I was so grateful when I finally got to shore 30 minutes later. Exhausted and still shaking from adrenaline, I finally let myself be mad at the stupid boaters and the stupid wind on this stupid lake.

I had to completely empty my boat and then tip it over to get the water out. And then repack. I paddled another 15 or 20 minutes and decided to pull off again to fasten my deck skirt on (a little late, I know). I continued to paddle hard into the wind, barely creeping along, until about 5:15pm when I got to a small public beach at Old Frontenac. I couldn't wait to pull over and jump out. My arms were aching, and I was done for the day. I paddled a little bit past the end of the public park, unloaded my boat and soaking wet gear, and found a place where I planned to set up my tent once the sun set.

I texted Jeff my situation and plan. He'd already arrived at our pre-determined camp site about two miles ahead. I just didn't have the energy or motivation to go those final two miles. Paddling 28 miles into the wind was enough.

I hauled some food and my computer bag to the one picnic table and had a nice hour-long talk on the phone with Leslie. The call really helped lighten my mood.

About 7pm Jeff texted, "It's still not too late to paddle over." My proposed campsite sucked, and there was still another couple of hours of daylight, so I decided to reload my canoe and took off, arriving on the beautiful sandy beach where Jeff was camped by 7:45pm. I was glad I did.

My equipment bag had gotten some water inside and my sleeping bag was wet. One food bag filled with water. But everything else was fine and still pretty dry.

It's now 9:20pm. My sleeping pad is wet. I have a few things hanging on the lines inside my tent. Super muggy!! And a thunderstorm is on the way.

What a day!!

We still have 15 miles to paddle on Lake Pepin tomorrow. And it'll be Sunday, so the boats will be out in force again, beginning around 10am, and it might be stormy. Hopefully the wind won't be as bad tomorrow. . . .

And the Wind and Waves Got Worse

August 18, 2019 - Day 23
Star Point Campsite to Wabasha, Minnesota – 21 miles
Total Mileage: 587.8

Lake Pepin . . . Aaaahhhhh!!!

Our little sandy campsite was pretty calm when we got up around 6am. The slight breeze was out of the north, which was good news. It switched as the weather forecast had predicted.

It rained and stormed overnight. Everything was wet and sandy, so it got packed up wet and sandy. The inside of my two cheapest dry bags also had an inch or two of water inside from yesterday's swamping. So, I emptied them out, and put my sleeping bag and Therm-a-Rest pad into extra garbage sacks before shoving them into their dry bags.

Coffee . . . Pop-Tarts . . . time to get paddling. I made a last-minute decision to forego my spray skirt and life jacket. Big mistake!!

As soon as we paddled around the end of the island we were camping on, the wind started to blow. It was at our backs and didn't seem bad at first, but within minutes of heading out into the open water, the wind picked up and the waves stared to pile up. After paddling only 10 minutes, I started to get nervous. We had decided to head about a mile across open water, taking a straight shot to a nearby peninsula. We were about a half-mile from the shore—too far—no spray skirt in waves that had grown to three feet—too high—deciding to head into open water to shorten the distance to the peninsula—too dumb. Day 23. When will I learn?

The waves were tossing my little canoe around. It was hard to steer. I focused my concentration on the nearest point of land, paddling as hard as I could before the next wave rolled over my stern causing it to fishtail. I was trying to keep my balance in the center of the boat as the building waves tossed me around.

I was constantly in fear of rolling over so far from shore. I eventually reached forward to grab my life jacket and carefully put it on and zipped it up while trying to keep my balance. A 13-foot open canoe in big waves on a huge lake feels like a cork just bobbing around. At one point, I heard Jeff yell from somewhere behind me, "Jon just put his life jacket on. This shit is getting real."

The next hour and a half were some white-knuckled paddling. I kept aiming for a peninsula that never seemed to get any closer.

Once around the back side of the peninsula, and out of the direct wind and waves, I landed my canoe and attached my fabric spray deck for the rest of the day.

The wind and the waves just kept getting worse. We decided to hug the shoreline, paddling only 15 to 20 yards from the rocky shore. The wind and waves steadily picked up over the next few hours and we got our asses kicked. The following and crisscrossing waves built from two feet to three feet to four feet tall. There were times when Jeff was ahead of me, and I could only see the top of his shoulders and his head above the waves. The rest of him and his boat were down in a trough.

My round-bottomed canoe was tossed around with every wave. The wind blew consistently at 20 to 25mph with gusts above that. And there was no sandy shore to pull off on, just jagged rocks and boulders. I felt way better with my spray deck and life jacket on but kept thinking we were pushing the envelope a bit too far.

We both really wanted to be off Lake Pepin. There were three spots along the way, where a small peninsula stuck out and we could tuck in behind it to take a little break. And then we'd paddle out into the four-foot waves again. It was exhilarating, exhausting, and a bit scary.

When those big waves came up from behind, my canoe would rise up and surf the top of the waves for a few seconds, with the stern fishtailing around. I was constantly pushing one rudder pedal all the way down and then immediately pushing the opposite pedal all the way down to flip the rudder in the opposite direction to counteract the rear of the boat whipping back and forth.

I tried paddling with my sunglasses on but felt like I could read the water better with them off. The direction of the wind and waves kept pushing me towards the rocky shore, and sometimes I'd be swept only a few yards from the huge rocks and would have to jam down my left rudder pedal and paddle like hell towards the center of the lake for thirty seconds or so to get a safe distance away. It was a constant and exhausting battle to stay upright.

We eventually got to the end of 20-mile-long Lake Pepin, and the lake narrowed to a wide river again. There was still wind in our faces, but the waves were a lot smaller as we paddled the last six miles into Wabasha. We'd decided pretty early on that we weren't going to make the full 26 miles we'd planned on going today. When we got to Wabasha, after six and a half hours of hard, non-stop paddling and sphincter clenching, we were both ready to be done.

We found the community campground just a block off the river, behind Slippery's Bar and Restaurant, and that was home for the night. Apparently portions of the movie Grumpy Old Men were filmed at Slippery's, and sure enough there were two very old and very grumpy looking old guys down on the river's edge with their fishing poles as we pulled up.

I felt kind of dejected about only going 21 miles and stopping around 2:15pm. But enough was enough.

We'll get back at it tomorrow.

On to Winona!

Locking Through

August 19, 2019 - Day 24
Wabasha, Minnesota, to Fountain City, Wisconsin – 30 miles
Total Mileage: 617.8

Today was originally the day that I was hoping we'd get all the way to Winona, Minnesota, to see Hanna (my niece) and Hunter (Hanna's boyfriend), but because of our big wind and wave day yesterday and only making it 21 miles, I knew we wouldn't make it all the way in one day. So, I texted Hanna that we'd shoot for Fountain City, which is seven or eight river miles before Winona.

We had two locks today. First, Lock #4. We could see a tug and nine barges sitting on the side of the river off in the distance about two miles before we got to the lock. It looked that they were just tied off and sitting there, but as we got about a mile away, black smoke started coming out of the tug's stack, and it started pushing the barges towards the lock. Poop.

Maybe 15 minutes earlier and we could've slipped ahead. So, we got to watch firsthand the lengthy process of a big barge and tug locking through. Interesting, but unfortunately it took an hour and a half. First, the connected raft of nine individual barges is very slowly and carefully eased into the lock, massive gates close, water drops, opposite gates open, a small support tug on the other side slowly eases the barges out, massive doors close, lock fills back up, upriver doors open again, tug pulls in, and then the same process as the tug locks through.

During the wait, I got out of my canoe and walked around to see if there was any way to portage around the lock and dam, but with only one set of portage wheels, it was way too long of a walk and not worth completely unloading, hauling our stuff and boats up a steep bank to a gravel road, and then portaging everything one at a time about a half-mile around the dam.

When we got to Lock #5, about 20 miles later, we again had a barge in front of us. I think the same one, so we had another hour-long wait.

There was a very irritated yacht owner anchored up and waiting to go through Lock #5 when we paddled up. I sat in my little canoe next to this big yacht listening to this guy complain as he sat beneath his canopy drinking cold beer with his wife while Jeff and I sat in our open boats in the glaring hot sun, sweating our asses off, hoping he'd offer us a beer.

We eventually paddled into the Fountain City (Wisconsin) boat harbor just before 6pm. I had just pulled my boat and trailer a couple of blocks to the Fountain Motel, where I'd reserved a room, when Hanna and Hunter drove up. So, they helped Jeff haul his bags from the docks to the motel.

We set up our soaking wet tents in the front lawn of the motel to dry them out while we drove to the Triangle Irish Pub and Restaurant and had the most amazing traditional, made-from-scratch Irish meals, all locally sourced and all made from the owner's Irish grandmother's original recipes. We had a really nice time with Hanna and Hunter, and then went back to the motel for showers, emails, journaling, and bedtime after a very long day.

Choose Your Poison

August 20, 2019 - Day 25
Fountain City, Wisconsin, to Dakota Island – 26 miles
Total Mileage: 643.8

We hauled our boats and gear back down to the river at the Fountain City boat docks and were paddling by 8am. Our two locks today, 5A and 6, were a breeze. The first one was maybe a 10-minute wait, and the second one, around mile 18 for the day, was opening its huge metal doors for us as we paddled up. Didn't even need to pull the cord to notify the lockmaster.

The guys at Lock #6 were super friendly. They told us how far into the lock chamber they wanted us to paddle, and then handed us each a rope to hang on to while the water level dropped.

The weather today was crazy!!

It started out cool, with a slight breeze out of the south, and overcast. Really nice paddling weather. Then, a couple of miles past Winona, we started to see big black clouds building behind us. The initial weather report said 20 percent chance of rain and thunderstorms from 1 to 3pm. When we noticed the dark clouds heading our way, I rechecked the weather and it had jumped to 80 percent.

We decided to get off the river before the storm hit rather than waiting until it started raining and blowing, which is what we've done in the past—wait for the storm to hit and then have no immediate place to pull off.

We stopped at a riverside trailer park that had a covered picnic area, grabbed our lunch bags and water bottles, and sat under the shelter checking emails and waiting for the storm to arrive. I suggested to Jeff, "Let's just sit here for a couple of hours and let the storm totally pass."

"Yup," he replied.

Well, after thirty minutes of waiting, and after eating a peanut butter, honey, and Chex mix tortilla, it hadn't started raining yet. We were both feeling antsy, so we decided to start paddling, this time with my spray deck attached!

Just as we pushed off, the rain started, which was quickly followed by a fierce 25+mph wind that seemed to come from the opposite direction, right into our faces. I was confounded that the storm was coming up from behind us, but the strong winds were blowing upriver towards us.

We paddled as hard as we could into the wind and mounting waves. I watched the nearby shoreline, and we were maybe traveling a half a mile her hour! Ridiculous!!

Paddle.

Paddle.

Paddle.

Driving rain.

Going nowhere.

Paddling in big waves in my little, heavily loaded boat makes me nervous. Having the spray deck fastened down helps my confidence quite a bit, but it is still unnerving to paddle a full boat in these conditions.

The storm only lasted 45 minutes or so, and then the wind died, and the hot sun and humidity kicked in for the rest of the afternoon. Not sure what was worse, the driving wind and rain or the heat and humidity.

We ended up camping on a beautiful sand island, created years ago by dredging the main river channel. There are hundreds of similar sand islands peppered throughout the Upper Mississippi from dredging over the decades to keep the river open to barge traffic. And more islands are being created and added to every day.

Highlights of today:

- Stopped in Winona and walked a few blocks for good coffee and quiche around 10am
- Stopped at a little fisherman's bar for a cold pop around 3pm, and ended up having an early dinner of beer and pizza
- Found the most awesome campsite of the trip so far
- After we unloaded our boats, I jumped in the river to wash off the sweat and sunscreen; Jeff took a dip too
- Had a great long video call with Leslie; she looked beautiful, and I can't wait to see her this coming Friday!!

A Typical Day

August 21, 2019 - Day 26
Dakota Island to a Mile and a Half South of Lock #8 – 28 miles
Total Mileage: 671.8

Thankfully, we had the wind at our back. That helped a lot today. I was just saying to Jeff that whether we paddle 25 miles, or 28, or 32 in a day, it is still a full day of long hard paddling. Today we had a slight breeze at our backs all day long, and paddled 28 miles, and it was a hard day of paddling even with the breeze helping us out. No matter the mileage, we are putting in eight, nine, 10 hours of paddling every day.

The hardest part of the day for me is almost always between miles eight and 15. The first few miles in the morning seem to go by pretty quickly. And then, after we get past the mileage half-way point for the day, there is a light at the end of the mileage tunnel that is motivating. But it is these middle miles, when there are still a lot more to go, that are mentally and emotionally the most difficult for me.

The other thing that is becoming a constant, now that we are 26 days into this trip, is that for the first 45 minutes of paddling in the morning, one or the other of my shoulders ache, and one or the other triceps aches, and it takes close to an hour to paddle through the ache until it stops hurting. Every day. Even though I'm sure that my arms are way stronger then when I started 26 days ago, they still ache every single morning. And ache during the night when I sleep on my side.

We were packed up and paddling by 7:45am. And by 8am, I was already dousing my Packer cap in the water and putting it back on my head to cool down. It never really cooled off last night, so it got hot earlier than usual this morning.

We're still paddling past lots of bluffs on the Minnesota side of the river. Seems like some good rock-climbing spots. I can't believe that after 672 miles on this river we are still in Minnesota. Or at least Minnesota is on the west side of the river. I think tomorrow we officially trade in Minnesota for Iowa. Finally.

I stopped on a sand bar after about 12 miles to stretch my legs a little and wait for Jeff to catch up. Around 19 miles we stopped at a public boat launch and campground to have lunch. I had another lame tortilla with buffalo sauce flavored tuna, two old cheese sticks that I've been carrying with me for almost four weeks, and some pulverized Chex mix. I'm officially sick of my lunch options!

After an unending seven or eight miles of post-lunch paddling on a river that has widened to almost two miles and become more like a lake, we stopped in the little Wisconsin town of Genoa, right before Lock #8. The town's boat launch was an overgrown sludge pond that smelled like sewage. It looked like no one had used it in years. We tied off our boats and walked into downtown Genoa, which consisted of a three-block-long main street with three small bars and a motel that looked like an extra-long travel trailer.

We chose the Big River Bar. Jeff liked it because it had the word "river" in the name. A typical small town Wisconsin bar with an Old Style beer sign hanging out front and a few old-timers inside. There was a couple who said they'd passed us on the river in their boat earlier today. They were super interested in our trip and asked all kinds of questions. They couldn't believe we'd paddled almost 700 miles.

We had a crappy but cold beer and commenced charging our phones and my laptop. I finished an hour or so of work emails and office work, and then we each ordered the "special" which sounded way better than it was: pulled pork sandwich on white bread, canned corn swimming in a jalapeño cheese sauce (super weird), coleslaw, and battered fries. The fries were the best part.

Around 5pm we packed up our stuff, got back in our boats at the sewage depository that doubled as the boat launch, and paddled about 200 yards to Lock #8, and then another mile and a half to a beautiful sandy beach with flat tent spots.

It's 7:42pm. The sun is almost down, and things are cooling off. We have a nice fire going, a little coffee mug of Jameson whiskey for each of us, and some DEET to keep the biting flies and mosquitoes at bay.

I studied my Army Corps maps tonight to assess where we can be by Friday night (today is Wednesday) when Leslie comes for a visit. We're looking at paddling 45 to 55 miles over the next two days, with a place to stop that isn't too far from a town with a hotel. It looks like it might be Prairie du Chien, Wisconsin. I'm just bummed that Leslie has to drive almost eight hours one way to get here!!

Mannequins Are Creepy

August 22, 2019 - Day 27
A Bit South of Lock #8 to Lynxville, Wisconsin – 27 miles
Total Mileage: 698.8

Good day.

It was another dewy night, so my tent fly is soaked. We were totally fogged in when we woke up. We couldn't see the other side of the river. Coffee and a bowl of granola with dried cranberries along with a breakfast fruit bar.

We left around 7:30am and paddled the first five miles in dense fog. I could only see 30 or 40 yards ahead. I just stayed close to the shore, which I could barely see some of the time. It was chilly. About 50 degrees. I really enjoyed having to stay focused and not get lost in the fog. It made the first five miles go by quickly.

At 15 miles, we got to the cute little town of Lansing, Iowa (we crossed from Minnesota to Iowa yesterday). In Lansing, we stopped at Shep's Riverside Bar at 11am, tying off our boats at the bar's personal dock. So, we'd covered 15 miles in three and a half hours, averaging 14-minute miles, or 4.3mph! That's a speed record for us!

Since we were making such great time, we stopped for two hours and charged our devices. I did some computer work. It was a nice break. On days like today, when we don't want to get to our campsite too early in the day and just sit around in the hot sun, it's nice to take a paddle break along the way.

The next 10 to 11 miles were tough.

The river broadened to over two miles wide. So, in essence, it was a big, long lake. No current. And things we could see in the distance were miles and miles away. Everything seemed to take forever to get to. The sky was cloudless, and even though it was only 75 degrees, it felt really hot. I probably dunked my hat in the water 100 times.

This kind of paddling is hard for me . . . three hours on a big, long lake where it feels like no progress is being made.

We decided to stop in Lynxville, a tiny Wisconsin railroad town. The map showed a "trailer park and campground." The place was pretty run down with nowhere really to set up our tents. However, there was a bar named Hoochies II.

We walked into Hoochies to ask about a camp spot.

"You want to camp? Here? In tents?"

That's what the lady behind the bar asked.

To get to Hoochies, we had to paddle off the river into a man-made marina-type spot. Most of the boats looked like they hadn't been moved in many years. The water in the marina was full of thick green sludge. The place was sort of creepy.

There was an old motel next to Hoochies, so we opted for a room with two queen beds for $60. We are the only ones staying here.

I did a little laundry in the shower and hung it all out on the railing by the front door of our room. I also hung out my wet tent and soaking rain fly.

Pizza at Hoochies. Got a little computer work done and watched most of the Packer's third pre-season game.

Tomorrow we only have a 15- to 16-mile paddle to Prairie du Chien. I'd sure like to go further since Leslie won't get here until 8pm or so. But it's a big town and is the best and easiest place to stop and meet up with Leslie, Jeff's dad Daryl, and his wife Bonnie.

Hopefully we can get our laundry done, restock a little food, and maybe even squeeze in a nap.

I can't wait to see Leslie!! It's been a month.

We are just over a mile short of 700 for the trip. After our short day on Friday, and another planned short day on Saturday (since our families are here), we need to crank out several back-to-back 30-plus days.

We can do it!!

Almost forgot. As we hauled loads of gear between the Hoochies motel, bar, and our boats, we passed a couple of cabins that had an array of very creepy mannequins both on the porch and inside the windows, wearing weird clothing, and posed in creepy positions, like hanging from the porch ceiling, and embracing each other.

Cabins inhabited by mannequins . . . think about it.

Dog's Meadow

August 23, 2019 - Day 28
Lynxville, Wisconsin, to Prairie du Chien, Wisconsin – 18 miles
Total Mileage: 716.8

Today I got to see Leslie after 29 long days!!

It's been a while since we were apart for over four weeks. I was really excited and had lots of energy to paddle today. We only had 18 miles to Prairie du Chien (which apparently means "dog's meadow" in French) and have already reserved two hotel rooms for the night. One for us, and the other for Jeff and his dad and stepmom.

We took off from creepy-ville around 7:45am and got to our take-out spot by a bridge, only two blocks from our hotel, by noon. So, we paddled over 4.5mph! That's motivated paddling.

Daryl and Bonnie weren't coming until 6pm, and Leslie not until seven or so, so we had plenty of time for laundry, showers, and some food shopping at the local Walgreen's. The hotel let us store our boats and extra bags in a nearby garage, so we didn't have to chain up our boats and haul all of our stuff into the rooms.

Leslie texted right before she arrived, so I was out in the parking lot waiting for her, and jumped up and down when she drove up, looking more beautiful than ever. She brought several new dry bags that I'd ordered, and the extra-large map case that Chris sent. Some of my bigger dry bags have been leaking. Just old and worn out.

We all had dinner at a local pub and caught up on family news and trip details. Everyone is staying an extra day so we worked out a spot where we could be picked up at the end of the day. Leslie is planning to go for a run and get some work done while Daryl and Bonnie do some sightseeing.

Our nightcap was a trip to the complimentary hotel ice cream and cookie bar. Nice!!

A Restaurant That Only Serves Brats
August 24, 2019 - Day 29
Prairie du Chien, Wisconsin, to Gutenberg, Iowa – 21.2 miles
Total Mileage: 738

Jeff and I both got up at our normal 6am. Leslie agreed that the sooner we started, the sooner we'd finish for the day.

Everyone helped us walk our boats back to the bridge where we stopped yesterday. I always think it must be a funny scene, two guys pulling canoes on wheels through town. We paddled with empty canoes, so went a little faster than usual. It was fun knowing that at the end of today's paddle Leslie would be there waiting.

It was a quick 22 miles. We were off by 7:30am and finished by 12:45pm. Five hours to paddle 22 miles. And it also got really windy the last few miles.

Since it was a Saturday, there were lots of fishing boats and power boats zipping up and down the river. It brought back bad memories of getting swamped on Lake Pepin. My boat sits up higher on the water when its empty, and bobs around more in the waves. So, it felt less stable for most of the day.

Right as we were nearing the lock at Gutenberg, a big barge edged ahead of us. We were really battling the waves and wind at that point, made especially nerve-racking when barges pass.

Since we knew that the barge would take at least an hour and a half to lock through, we opted to pull our boats out at a park just upriver of the dam. And right as we pulled up to the side, Leslie showed up, waving.

Leslie and I put my canoe on the car and took off for Marquette, Iowa. Leslie had stopped there earlier today on her way to Gutenberg. We had a late lunch at a riverside place, McGregor's Beer and Bratz Garden, that only served, well, cold beer and brats. We sat by the river and got caught up on life and news from the past 29 days. After lunch we went to a casino and won $200!! Afterwards, we checked into our cute little Airbnb in Gutenberg. It was a second story studio in an old red brick building. I loved it. But no internet. I called the out-of-town owner, and he had me down in the basement looking for boxes and wires to wiggle and check. To no avail.

Daryl suggested dinner in a nearby town at a grubby looking biker bar. It ended up serving amazing food. Fun night. Great second day with Leslie. I already miss her, and she's still here.

Canned Hamburger and Velveeta Cheese

August 25, 2019 - Day 30
Gutenberg, Iowa, to Mud Lake Campground – 26 miles
Total Mileage: 764

It was a blustery day!

Started with wind in our faces . . . ended with wind in our faces.

Leslie, Daryl, and Bonnie saw us off at the boat landing across the street from our Airbnb. It was hard to say goodbye to Leslie. Part of me wanted to just keep my stuff in the car and drive eight hours back home with her.

We were on the water by 7:45am. Daryl filmed our departure. The wind was blowing 10 to 20mph out of the south, according to the

NOAA weather report I was listening to. And we felt every mile per hour.

We had a couple chances to duck behind some islands to try to get out of the wind, but it seemed like no matter what we tried, the wind kept hitting us head on. It was exhausting.

Around mile eight, I pulled over at a boat landing in the little town of Cassville, Wisconsin. Jeff pulled up 15 to 20 minutes later. I just needed to lay in the grass for a while to rest a bit.

We sat under a picnic shelter and had some of Bonnie's awesome zucchini bread, and the rest of my cold coffee from earlier this morning.

The predicted 20 percent chance of scattered showers struck again and again. I think it rained a total of six times on and off during the day. I never put my raincoat on, always thinking it would just be a light sprinkle, which it was sometimes, and other times it rained hard and blew for half an hour or so. I don't mind getting a little wet, especially on days like today. But today I got a little chilled. Just as the wind dried me out, it would start to rain again.

We stopped again around mile 16 in a small private boat harbor. It was just a bunch of travel trailers, a few boats, and a little picnic pavilion where we sat and had our lunch of tortillas, canned hamburger (seriously), and Velveeta cheese. Jeff provided lunch and was super excited about it. I thought it was gross, but since he was providing it, and it was something different from the usual, I kept my comments to myself.

The wind died down for miles 16 to 23 or so, and then really kicked into high gear for the last two or three miles. It was blowing so hard that I'd paddle as hard as I could and only crept along maybe a half-mile an hour. I hate that kind of almost pointless paddling.

By the time we pulled into Mud Lake Park around 4pm, we were both exhausted. I could have gone to bed and slept all night, right then and there. We were originally planning to stop a few miles back at an Army Corps campground, but realized that if we pushed another few miles, we could get to this county campground that was right next to The View Riverfront Bar and Grill. Motivators like that can make a person do almost anything.

So, we set up camp and have been sitting at the bar and grill for the past three hours, I've been catching up on emails, reading the news, talking with Leslie and Seth, and doing some office work.

It's almost 8pm and I can barely keep my eyes open.

Tomorrow is supposed to be stormy, so we have a couple of contingency plans in place.

Who Cares?

August 26, 2019 - Day 31
Mud Lake Campground to Dubuque, Iowa – 11 miles
Total Mileage: 775

Today was sort of a pre-destined bust. If we'd started the day by saying we were taking a weather day off, and then decided a few hours later to go ahead and paddle 11 miles, we'd have felt like heroes. But instead, we had the idea of going to the Dubuque Marina and then hopefully another 16 or 17 miles and ended up staying at the marina. So, now I kind of feel like I wimped out, or backed down from the challenge we'd set. I know that no one else cares, and I don't really care in the big scheme, but still. . . .

When we went to bed last night, and when we woke up this morning, the weather radio said 10 to 15mph winds out of the south and east (why isn't it ever of the north, and at our backs?), intermittent rain in the morning, and increasing wind, rain, and a chance of thunderstorms in the afternoon and evening.

So, the plan we made last night was to not leave this morning if we woke up to a rain shower, and then paddle to Dubuque (11 miles) if the wind and rain allowed. And then pull off at the Dubuque Marina and decide from there.

Well.

We got up.

No rain.

Just cloudy and breezy.

Took down the tents lickety-split before it started to rain.

Brewed up coffee.

Downed two instant oatmeal packets.

Attached my canoe to the trailer wheels, loaded it up, and wheeled 'er in.

There was a bit of a breeze out of the east, off to our left, but not bad. It was a seven-mile paddle to General Zebulon Pike Lock and Dam #11. We were passed by a huge fake paddle wheel boat that I found out later goes on 15-to-22-day river trips on the Mississippi between Red Wing, Minnesota, and New Orleans.

The fake paddle wheeler locked though ahead of us, but we only had to wait 20 minutes or so. Then it was a four-mile paddle to the city of Dubuque and the city's big marina.

Almost no rain, not much of a breeze, and no lightning, yet. But the skies looked dark and ominous.

So, we tied off right next to the big fake paddle wheel boat that had passed us and was disgorging its passengers. Lots of stares at our tiny, heavily loaded boats on what is becoming a bigger and wider river.

We got coffee at a visitor center and pulled out our maps to think through the next few days to help decide whether we should push on into the ominous looking weather. Jeff has four and a half paddling days after today, before he gets picked up and ends this first half or our Mississippi River adventure.

We came up with two scenarios. One, we paddle another 17 to 18 miles today (because there is really nothing between here and the next 18 miles). Two, we spend the rest of the day here in Dubuque, just goofing off.

Well, goofing off won. Hence my little bit of guilt that we only paddled 11 miles today. But as my dad has been saying in his later years, "Who cares?"

We unloaded our boats and tied them up at the marina for a $25 fee. We found a nice hotel within walking distance for $91, grabbed a few things from our boats, and headed over. The rest of the day consisted of going to Subway, and me playing slot machines at a casino and winning $513!!!! I dd some office work on my computer for a few hours and walked with Jeff to a great beer place. I had a Cuban sandwich with amazing homemade mac and cheese and treated Jeff to dinner with my gambling winnings.

Now we're sitting at the hotel bar having a drink and journaling. A fun day. It's only 6:45pm and I can barely keep my eyes open.

Oh, we also took and Uber to an O'Reilly's Auto Parts store so I could buy a sponge to soak up water that collects in the bottom of my canoe. I think I have a very slow leak.

Tomorrow we'll get back to some serious paddling and miles. It's weird how I feel guilty about only going 11 miles today and have a hard time accepting that today was a fun day with my cousin, and that's good enough. Who cares?

The River Chaplain

August 27, 2019 - Day 32
Dubuque, Iowa, to Pleasant Creek Recreation Area – 26½ miles
Total Mileage: 801.5

The campsite we are at tonight, Pleasant Creek Recreation Area "Primitive" Campground, was under three or four feet of water just a month ago, which would mean a river that is 10 feet higher than the one we are paddling right now. This entire campground was underwater for over a month due to river flooding.

We heard the same thing a few nights ago when we were camping by Mud Lake. The entire camping area had been five feet underwater, and people had their fishing boats floating around where the campground should have been, catching fish.

So, the Mississippi River flooding that we read a lot about earlier this year really devastated much of this area. We heard from someone today that in some places near Cape Girardeau, Missouri, further south, the river spread 20 miles inland!! Here at the campground, there is driftwood resting all the way to the top of the railroad tracks, 200 yards from the river.

The high wind warnings of gusts to 30mph that we'd heard in the weather reports last night and this morning never really materialized. There wasn't even a breeze until early afternoon, and then it was only maybe 10mph and mostly at our backs, pushing us along. So that was awesome.

We paddled about 27 miles today. It felt good to get a long day of paddling in. Around mile six, we stopped at a little hillbilly marina for a cold drink. They had a tiny 10-foot by 10-foot stage for live music on the weekends, and a nice little campground. No one was there at the marina, but thankfully there was a pop machine that dispensed some ice-cold Mountain Dews.

Around mile 15 or 16, we pulled over to the west shore to get out of the sun and under some riverbank trees for lunch. I ate leftover brisket and the other half of my Cuban sandwich that was left over from last night's dinner at the marina in Dubuque, and a couple of pieces of Bonnie's toffee.

Around mile 21 or 22, we stopped in the town of Bellevue, Iowa. Another sleepy little riverside town. Jeff bought two more pairs of sunglasses. Pairs #4 and #5. The others have been broken or lost.

Back down by the river, I saw a guy walk up to where we had tied up our boats, stand there for a minute, and then get down on his hands and knees and start to open my yellow dry bag full of food sitting in my canoe. WTF? I yelled, "Hey, can I help you?" And then I yelled again as I got closer. Pretty ballsy to be digging through our stuff in broad daylight at a public boat ramp.

It turned out that he was fishing offshore and got his lure caught on one of my dry bags, and he was trying to get it unhooked. Funny. I told him about getting two of three barbs on a treble hook caught on my underarm just before the start of this trip and needing to go to the hospital in Calumet to get the hooks cut out.

We also met two old guys who were hauling their aluminum skiff out of the river after a morning of fishing. They asked about our trip and told us a few stories about how nice Iowans are . . . the nicest folks in the country. I'll bet any money they were both Trump supporters which made the conversation another eye-opener for me.

We locked through Lock #12 just above Bellevue. There was a tug with twelve barges ahead of us, but I called the lockmaster and he let us jump ahead and lock through without have to wait two hours.

A bit before we got to the lock, we paddled up behind another kayaker. From a distance, I thought it was Bob, our paddling buddy from earlier in the trip. This guy paddled sort of erratically, just like Bob did. It turned out that Steven, the first river paddler we've run into other than Bob over the past 32 days, started at Itasca State Park three weeks before we started!

I knew right off that something wasn't quite right with Steven. He was paddling a plastic lime-green, sit-on-top kayak, with a big duffel bag strapped to the back. No waterproof dry bags! He also had some miscellaneous items strewn about on the front of his boat. From a ways back I could see that Steven's boat was zigzagging back and forth, making very little progress due to his erratic paddling technique and the fact that his Walmart kayak didn't have a rudder.

Steven told us that his Appalachian Trail hiker nickname was "Son-Shine Hiker," with a special emphasis on the "Son," and that his Mississippi River paddler nickname is "River Chaplain" since he's been ministering to homeless people along the river. Yup. That's really what he said.

Steven also told us that yesterday in Dubuque he was camped across the river from the city, at a spot that had no road access, and that someone stole his canoe. So, he hitchhiked to town (remember the "no road" part of the story?) and met someone who took him to Walmart (I knew it!) and bought this $200 plastic sit-on-top, lime-green kayak so that he could continue his trip down the Mississippi.

Steven continued on, telling us that on the first day of his trip, back in Itasca State Park, he was staying at the hostel lodging facility at the park. He was starting his trip with two inflatable kayaks that were tied together. One for him, and the second one to tow his gear. He said something about paddling off from the hostel, thinking that he was just taking a short practice paddle, and ending up paddling all the way to Bemidji, three days away, without any of his gear or clothes. He said that for three nights he slept with no tent, or sleeping bag, or any of his stuff. He only brought his food with him. He said he'd brought his food along on this short practice paddle so that animals wouldn't get into it at the hostel.

Is any of this making sense?

I have been sitting by this nice campfire that Jeff built tonight, hoping and praying that the River Chaplain doesn't come paddling around the corner.

It's a beautiful, calm night. Getting chilly. Seems like we've been on the river forever . . . like Itasca and Day #1 were months and months ago. This is a long, long river, and 2,400 miles is a long, long way to paddle. But like any big extended adventure, it's one day at a time. And here we are, over 800 miles already. I plan to make it to Hannibal, Missouri, for sure. That's been the spot I've always had in mind for the end of this first leg of the trip.

I was saying to Jeff earlier today that he needs to get ready for no one back home being able to understand this trip, or the amount of effort required, or what any of it has been like. Words can't describe the hundreds of emotions and experiences and sights and sounds that make up this river.

Ten more days. Part of me wants to just keep going.

Bushwhack Campsite
August 28, 2019 - Day 33

Pleasant Creek Recreation Area to Horseshoe Island – 27 miles
Total Mileage: 828

Lots and lots of eagles today. Seemed like I saw an eagle every few minutes. I always keep my eyes open for eagle feathers. I've found three so far, but none today.

About halfway through the day, we took a side channel to get off the big wide river. Just looking for a change of pace. It looked fine on the map but ended up being one sand bar after another. We had to get out of our boats several times and pull them along through the shallow water using the bow line. I enjoyed getting out and walking, so it offered some nice breaks.

There was a stiff wind, but it was out of the west, so we pretty much hugged the right Iowa shore all day.

After 15 miles, we stopped in Sabula for lunch. Sabula is a tiny little town on the Illinois side that is actually an island in the middle of the river. Sabula was a rundown little place with one café, the Sand Bar and Grill. We tied off our boats at the community boat ramp and walked a few blocks to the café. Inside were six or seven old-timers who all gave us a good, long look as we came in the door. Great food. Super cheap. Bottomless Mountain Dew. I checked emails and we charged our devices before heading out.

By mile 24, the lake opened up to 3 miles wide—the widest point of the entire Mississippi River. We decided to stop at the first sandy spot we found. But the sandy spots seemed to have disappeared. We paddled and paddled, and the shorelines of the several islands that we passed were surrounded by a dense 100- to 200-foot buffer of lily pads and river plants. We couldn't get anywhere near land. Even the sections of open water between the islands were completely clogged with aquatic plants.

Finally, about 5:30pm, 10 hours after we'd shoved off this morning, we got to the last island before a four-mile section of open water: Horseshoe Island, named by us due to the shape. It was completely surrounded by a 200-foot-wide barrier of water plants. We paddled as hard as we could, plowing through the weeds to a point maybe 20 yards from shore where we had to jump out and pull our boats through thick weeds the rest of the way to land.

It was a densely treed island, but we found a spot that was barely big enough to set up two tents and start a small fire. So, with no other options, we unloaded our boats. It was buggy, and cramped, but we were both happy to be somewhere.

Sticky, sweaty, tired, and camping at a bushwhack site in the trees with no breeze.

We had miso soup for dinner, and I made a peanut butter and jelly tortilla. Jeff made a nice fire, and we finally drank the three warm Surly brand cans of beer I've been hauling with me the last two weeks. Sort of a lame end to the day.

Just Water Over the Dam

August 29, 2019 - Day 34
Horseshoe Island to Camanche, Iowa – 18 miles
Total Mileage: 846

The wind was coming out of the south and southwest a steady 20mph with gusts above 30mph for sure. All day! Non-stop!! We'd originally planned to paddle to the little town of Princeton, Iowa, something like 27 miles. But there was no way.

We packed up from our buggy bushwhack site and were both dragging our boats back through the thick aquatic plants by 7:15am. There was already some wind, and our little horseshoe-shaped island was sitting out in the middle of a 3-mile-wide section of the river. We planned to paddle the four miles to Lock #12, right down the middle of this lake-like river section, but early on the wind was causing big waves that made me nervous. Paddling straight into the wind and waves would've easily taken us over two hours to get to the lock, and in that amount of time, the waves could build to something more dangerous, and we'd be at least a mile from either shore. Floating outside my canoe, in big waves and over a mile from shore, didn't seem like the right choice.

So, I told Jeff that I wanted to head to the nearest shore on the Iowa side, and just paddle along the shore. It ended up adding about two miles to get to the dam, and an extra hour of paddling into the wind, but it was the smart thing to do. I felt a lot more comfortable once we got closer to the shoreline. But for the next three hours it was paddle, paddle, paddle, rest, blow backwards, paddle, paddle, paddle. Exhausting. Jeff was behind me somewhere, but his kayak handles so much better in choppy waves, so I wasn't worried about him being out of sight. I just wanted to get to the lock and out of the wind and choppy water.

As I got to about a half-mile before the dam, which, like the others, is a series of 11 or 12 independent metal gates that can be lifted and lowered depending on the water level, each a few hundred yards wide, I started to cut back to the left side of the river, to the Illinois side. The lock we'd paddle through was on the far-left side, which is only the second or third time of these first twelve that the lock has been on the left side of the dam. I paddled hard to cut across the front of the 12 separate dam gates.

As I got partway across the front of the dam, I saw two lock staff in their neon green t-shirts walk out on the concrete lock barrier that creates the boat channel for the lock. They waved at me. I watched (and paddled) as one guy got into the speed boat that is always hanging from a hoist, ready to be dropped into the water in case of a boating emergency.

I kept paddling hard, and when I got close enough to hear him, he yelled, "What are you doing?" "My buddy and I want to lock through," I yelled back. He proceeded to tell me in no uncertain terms, that I should never have gotten that close to the huge dam gates, and that the rule is to stay at least 1,000 feet upriver from the dam. I had no idea. He said I could have easily been swept over the dam, and that a month ago the river current was so strong where I had just been paddling that I would have been swept over for sure.

I felt embarrassed and a little scared and apologized several times. The lock-guy said, "I'm not yelling at you. But I was getting ready to drop my boat in the water for a rescue." He asked if I had felt the current and undertow that was sucking me towards the dam. But the truth was that the wind was blowing so hard in the opposite direction that I never felt the pull.

I tried to call Jeff to tell him to stay more upriver, but he didn't answer his phone. So, the lock-guy stayed near his boat and watched until Jeff made it over safely to where I was. I felt stupid, but glad I'd learned my lesson that way and not a more dangerous way.

We locked right through and paddled another few miles to a boat marina in Clinton, at about mile 10 for the day. The wind was just blasting us, and we were only going maybe a half-mile per hour. Just barely moving. We tied off and went in to get out of the wind and get an early lunch. I'd burned through my oatmeal long ago.

Back on the water, we slowly powered into the wind. It was getting stronger, and the forecast was for winds above 30mph and gusting to 60mph!! Plus, thundershowers and a chance of hail and maybe a tornado later today. Shit!

We took a side channel to get away from the big barges and ended up in a wind tunnel where the wind was funneled directly into our faces for four miles. Big waves. At one point I pulled off to put my spray deck and life jacket on.

We were exhausted.

We made very slow progress to the little town of Camanche at our 18-mile point and called it quits for the day. Our boats are tied up at little Comanche Marina and the owner is letting us camp here in the grassy yard for $10. We'll head over to a local bar in a few minutes and get a pizza and watch some sports. Very glad to be off the river. Looks like the most severe weather missed us.

A really nice local guy, Mike, just stopped over where we're sitting at the marina to say hello. Super friendly. Reminded me of Paul Stine, my son's father-in-law. Mike said, "If you ever write a book about your trip, mention that you met a really nice guy in Comanche."

We're sitting in Hide's Bar and enjoying a frozen pizza dinner and cold beer and talking with a few locals about our trip. Camanche is a very small town, and everyone in the bar knows we're not from here. As we were finishing up our beers, a guy walked over and offered to bring his little trailer to where our tents are set up at the marina so we can sleep dry tonight. It is supposed to rain hard tonight.

We said, "No, thanks," to Ray, and then his wife Erin came over and gave us the hard-sell. Not being two guys to look a gift trailer in the mouth, we climbed into their truck and headed to their house to hook up the trailer. After not hooking up the trailer correctly, pulling out of the driveway, and having the trailer pop off the hitch and drag on the ground for several feet, we eventually made it back to the marina. Ray was very excited to show us his custom-made trailer. This trailer was tiny with only a large mattress inside. Like a tear-drop camper, but it was complete with an incredible sound system, flat screen TV, and DVD player. Ray stayed to show us how everything worked and seemed so happy to get us out of sleeping in our tents after 34 nights.

The trailer might have been perfect for Ray and Erin but would have been VERY cozy for two grown-ass cousins, so I offered the trailer to Jeff, and gladly crawled into my tent. It was after 11pm and way past our bedtimes.

Last Day with Jeff . . . For Now

August 30, 2019 - Day 35
Camanche, Iowa, to Davenport, Iowa – 29 miles
Total Mileage: 875

Our last day paddling together.

35 days!

It's hard to believe, and to accept, that this part of the trip is over.

We both agreed to not really dwell on it today. We figured we'll have the rest of our lives to process the trip and how it went. Plus, we had 29 miles to paddle today, so the trip wasn't quite over for either of us.

The predicted stormy weather never came last night. I slept in my tent and Jeff was in Ray's little camper. We both slept great. It was funny seeing Jeff crawl out of that tiny camper door.

The breeze was coming out of the north, so that was an awesome way to start the day.

Coffee and some hot oatmeal.

Packing up dry tents is always nice. I had to poop and found a Porta-Potty (after some serious and increasingly urgent hunting around) that was a few blocks away at a construction site. I was getting a little worried since the coffee had already gotten my gears turning.

It was a long paddle day. We started with a little current and breeze, but it totally petered out after just two hours. The guy at the lock we passed through (Lock #13) didn't even toss us a rope to hold on to. So, we just floated around in the middle as the water lowered only 12 feet or so.

In Davenport, we pulled into the Lindsay Park Marina, which turned out to be private. Jeff, the staff person on duty, was super nice and enamored with our trip. He let us store our boats and gear for free and kept saying that we were his guests and that we could help ourselves to the bathroom, showers, and snacks in the members-only lounge. A few of the yacht owners asked about our trip as well as we unloaded or boats. The further south we get, the more amazed people seem to be . . . 35 straight days of paddling . . . 875 miles!

We Ubered to a hotel, showered, and ventured out for a couple of celebratory beers on Jeff's last day on the river. The two nearby breweries got flooded out along with the rest of the Davenport businesses that were within three to four blocks of the river. So, both brew pubs were closed. All of these businesses had six feet of water in them from the river flooding in June, only two months ago. Pretty wild.

We ended up drinking mead flights at a meadery(?) and talking with Rachel and her husband, who run the place. Jeff got brats at a nearby outdoor restaurant, which we ate in the mead place. The brats were followed by some bottled beer at the meadery, then mixed drinks at the brat place down the street, and we staggered back to our room at 11pm.

Super fun.

We laughed and reminisced all evening.

Life Without My Brother

August 31, 2019 - Day 36
Davenport, Iowa, to Muscatine, Iowa – 29 miles
Total Mileage: 904

Phew.

I'm pooped.

29 miles.

It was a perfect paddling day. Low 70s. Cloudy all day. Very slight breeze at my back, sort of.

But 29 miles is a long way to paddle on this big river with basically no current. I left at 7:45am. Paddled about one mile. And then sat at Lock #14 for about 45 minutes before I was let through.

And I pulled out in Muscatine at 5pm. So, a long day.

The biggest difference today, my first day without Jeff, was that I don't have anyone to make decisions with, bounce options around, decide which way to go or where and when to pull off and eat, decipher what the weather and wind is going to do. So, the first hour or so I actually felt a little bit anxious about being on the river alone. But then I just paddled and paddled and settled into myself.

On these 30-ish-mile days, the first 10 or 12 go by easily, and the last eight or so seem to go by quickly. But the middle 10 to 12 miles just really drag on. It felt like I was paddling hard and barely moving. Like the water had turned to molasses. I think that first thing in the morning I'm ready and raring to go . . . to start a new day and see new things. And then during those last miles of the day, I'm thinking about how nice it will be to stop in an hour or two and be done for the day. But during the middle section of long days, I start feeling tired; I know I haven't gone very far yet, and I still have a long way to go.

At about mile 11, I got to the little town of Buffalo, Iowa. So, I stopped and had an early lunch at the only place to eat in town, Judy's Barge Inn.

I had a headache and felt sort of queasy all morning from drinking too much last night. I just felt sort of sick all day. I eventually paddled the alcohol out of my system, and it helped to get some food in my stomach. I had an awesome BLT and a salad. I'd just had coffee and a scone for breakfast at the hotel, so I needed to add something a little more substantial.

Man, it was hard to get going this morning. And it felt weird to be the only one packing up. Jeff took the Lyft ride with me back to the marina and helped me load my boat. I felt sad to be leaving Jeff behind, and I know he wished that I wasn't paddling on without him. But I need to see what it feels like to paddle all day, for several days, by myself, to see if it's something I want to do, to finish up this river by myself later this fall.

Jeff and I have really become great friends over the past 35 days. That's been the best part of this trip. Jeff is a fantastic travel partner!

I locked through #15 late in the day, just above Muscatine. I paddled right in. But coming out the other side of the lock was something else. When the massive lock doors opened, there were huge choppy waves everywhere. Like three- to four-foot-tall waves all swirling around in different directions, colliding with each other and bounding off the concrete walls of the lock chute. The waves must've been caused by water being let through the dam, and then backing up towards the lock doors.

I was really nervous paddling into these crazy waves, as I slowly picked my way through the turbulence from one wave to the next. First a big one coming from the left, then one quickly from the right, and then some reverb waves crashing from behind. I was worried about capsizing, and of course I didn't have my spray deck fastened down since I wasn't expecting these waves.

What I thought was a private marina in Muscatine turned out to be just a public dock with no place to safely store my boat. So, its sitting upside down in front of the Merrill Hotel in downtown Muscatine, where I'm staying.

Laundry is in the dryer. I'm having dinner in my room. And Day #1 of paddling alone is in the books. I'm exhausted!!

Paddling Alone

September 1, 2019 - Day 37
Muscatine, Iowa, to New Boston, Illinois – 23 miles
Total Mileage: 927

I was in a groove today.

Not sure what it was, or what was different, but paddling today almost felt like a meditation.

Most of the way, the river was like glass.

Paddling just felt like the thing I should be doing today.

No wishing I was somewhere else or wanting the paddling to be over with.

Yesterday, I thought I might take today completely off. Then I decided I'd do some work in the morning at my hotel, and then take off around noon and only paddle 13 miles and camp somewhere along the shore.

I slept in, sort of, until 7am or so. Took a leisurely shower, and then took my laptop to the lobby and had a cup of amazing espresso while I got some work done.

I had a nice omelet for breakfast, did some more work, and then packed up and loaded my canoe in front of the hotel. I pushed off at 11:20am.

I got to my 13-mile goal, and it was sunny and warm, and I didn't feel like stopping, so six more miles got me to Lock #17, and four miles beyond the lock was New Boston, Illinois. I got here about 5:30pm. I only got out of my canoe once today, while waiting to lock through. So, six straight hours of paddling.

It felt good to have gone 23 miles instead of the 13 I was planning. There is a boat launch and campground right on the water, full of campers and holiday boaters. I attracted some attention when I was unloading my canoe, and a guy walked over and welcomed me to town, saying that he was a local Alderman, and that I could set up my tent anywhere I wanted.

There wasn't really a good spot to camp, other than a nice grassy patch next to a gazebo that seemed right in the middle of everything. While I was contemplating where to haul my bags and set up, another guy stopped to say hello. He said that he was a New Boston city employee and gave me the passcode to the campground bathroom: 0515. A shower before heading out tomorrow morning sounded great.

I set up my tent at the spot by the gazebo and a few picnic tables, and then walked a few blocks towards the little town to Rocky's Sturgeon Bay Bar and Grill. My Alderman buddy was there, and he recommended either the catfish or the wings. Of course, I got wings and a salad.

I talked with Leslie and then got out my maps to figure out where to go these next few days. Since I paddled 23 instead of 13 today, my previous paddling plan needed some alterations.

Traveling alone on the river is different to me than traveling alone in a third world country. I feel more vulnerable out on this big river alone.

I'm also finding that I'm more social without Jeff too. It forces me to talk with people and ask for help.

Options for stopping places are pretty limited the next few days. I need to be in Quincy, Illinois, by Friday night (five days from now). I had been thinking Saturday, but Leslie decided she can come on Friday, which is awesome. I may end up doing a 32-mile day on Friday; it would be an epic way to end my six weeks of paddling.

There are gnats or no-see-ums out tonight, while I sit here at a picnic table writing in my journal. I haven't had to use bug repellent in the past two weeks. Too cool at night for bugs. Maybe I'll take my shower tonight after all and wash off the bug dope before crawling into my bag. It's only 8:10pm, and I already need my headlamp to journal. Too early to go to sleep. But I'm too unmotivated to do any office work.

Okay. Shower it is. Another good day on the river.

Tomorrow is supposed to be hot and muggy.

Oquawka

September 2, 2019 - Day 38
New Boston, Illinois, to Burlington, Iowa – 29 miles
Total Mileage: 958

Today was a tough one.

Zen paddling yesterday, followed by "I am not into this at all" today.

The forecast was five to 10mph wind out of the south, so in my face all day long. And no real current to speak of. And mentally knowing I had 29 miles to go into the wind made it psychologically much more difficult.

I tried to break things up into smaller sections, like I usually do. Five miles to Keithville; 10 miles and I'll get out of my boat and stretch. Today the hardest stretch psychologically was mile eight through 15. Around mile 17, I stopped in a really weird named town of Oquawka, Illinois. I was starving, and really needed a break. The town was tiny, but I saw a little bar just a block up from the river.

I walked in, and immediately noticed five young men, all sporting the exact same weird bowl haircut, homemade button shirts, and the exact same black suspenders. All five were seated in a line at a table, drinking ice waters and waiting for their food. Rock music was blaring in the background. Television on some sports channel. Mennonites? Quakers? Amish? Pennsylvania Dutch? So weird to see them all sitting in a bar. Probably the only place in town to get lunch.

I texted Leslie, "Guy walks into a bar in Oquawka and sees five Mennonites. . . ."

My half-pound burger took forever to come out. Fries were old and stale. But at least I was off the water and out of my boat for an hour.

Back on the river at 1:30pm with 11 miles to go. They were 11 long, slow miles, paddling into the wind.

A real grind.

A mental test.

Just tried to keep a relaxed mind and keep paddling hard, stroke after stroke after stroke.

Lots of fishing and pleasure boats out today.

It's Labor Day.

Lots of people waving as they zipped by in their speed boats, throwing up huge wakes.

Plodding along.

I finally got to the Bluff Marina a little after 5pm. Nine hours after I started out this morning. As I was tying off on one of the floats so that I could walk up to the marina office to see if I could store my boat for the night, a guy yelled down from the marina deck above me, "Do you want to tie your boat up for the night? Just leave it there. It'll be fine."

Jon, the owner of Bluff Marina proceeded to bring me a wagon to haul all of my bags into his boat showroom for safe storage overnight. He brought me an ice-cold Busch Lite and suggested that I call the local casino for a free van ride to the hotel and casino.

Super friendly. And really appreciated after a long hard day.

To meet someone so willing to help after a day like today was really amazing.

Jon said that he was a city councilman for Burlington. He seemed very proud of his city.

I spent an hour or so in my hotel room watching U.S. Open tennis and sorting through my paddling options for the next few days. The wind is supposed to be 10 to 15mph with gusts to 25mph out of the south (BAD NEWS!!). Temperatures are supposed to be

88 degrees and humid, with a heat index of 96 degrees (REALLY BAD NEWS!!).

And I have a teleconference tomorrow afternoon at 3pm that I can't miss.

So, if I take off in the wind and don't get very far, I may not have cell service where I end up. Right now, I'm leaning towards taking tomorrow completely off the river, which means I likely won't make it to Quincy by Friday, which poses a different challenge. I've already changed my mind a few times tonight, so I'll just go to bed and check the weather in the morning and see how I feel.

Zero Day
September 3, 2019 - Day 39
Burlington, Iowa – 0 miles
Total Mileage: 958

First zero day of the trip.

Wind 15 to 20mph out of the south (into my face). Gusts to 30mph.

Chance of thunderstorms.

Temperatures at 90 degrees with heat index of 98 degrees.

High humidity.

Tons of computer work to do and a teleconference at 3pm.

Plenty of reasons to take the day off.

Human Contact
September 4, 2019 - Day 40

Burlington, Iowa, to Fort Madison, Iowa – 22 miles
Total Mileage: 980

My emotions go up and own throughout the day. I started the day totally ready to be back on the river. It felt weird to not paddle yesterday. Abnormal. So, I was really looking forward to getting back on the water. Jon, the marina owner met me to retrieve my stuff and send me off on a positive note.

It took several miles for my arms and shoulders to settle in and stop aching. I took my first round of Advil earlier than usual and by mile five it was starting to bother me that my shoulders, especially, were still aching. So, my mood started to flag. I surprised myself at how I could feel so up to start the day, and so down only two hours into the paddle.

The scenery was pretty, but monotonous.

At about mile 10, I took a four-mile-long side channel/slough and it seemed to go on and on forever. No landmarks to watch for. It just went on and on and on.

I finally got to a small rural boat ramp around mile 14 that I'd circled on the map for a possible break.

I got out of my boat. Brought my lunch bag and water and maps up to a log on the beach and walked around a little to stretch my legs. As I munched on an apple, PowerBar, and pretzel mix, three different guys pulled up.

One on a moped: "What's happening on the river today?"

A second guy was hauling building materials to his partially destroyed river cabin.

And a third guy who had two black labs with him that he'd brought out to practice fetching in the water to get his dogs ready for duck hunting season. We talked about the river, hunting, and my trip. He'd met a few other through-paddlers over the years and told me stories about how unprepared some of them were. He met two kayakers at this same spot a few weeks ago, and they were on Day 50 of their trip (I'm on Day 40). He wished me well and as I paddled off, I reflected on how that little bit of human interaction really lifted my spirits for a mile or so.

The last six miles to Fort Madison went by very slowly and my mood dipped yet again. A big barge came up behind me, and I watched as the train bridge in front of me spun around on a giant rotating platform to allow the barge to go through.

Chi, the hostess of the old red brick Kingsley Inn, met me at the public harbor and boat ramp and hauled my bags to the inn while I pulled my canoe on its trailer the six blocks.

The Kingsley is a very cool old Victorian-esque hotel. And Chi was super helpful.

I spent the rest of the day getting more office work done.

Watching Rafael Nadal in the U.S. Open.

Exhausted. (Me, not Rafael.)

Headache.

Need to get to sleep.

Only two more days on the river before I need to head home. Part of me doesn't want it to end. This lifestyle. This rhythm. Part of me is ready to be done.

1,000-Mile Mark
September 5, 2019 - Day 41
Fort Madison, Iowa, to Keokuk, Iowa – 22 miles
Total Mileage: 1,002

I hit the 1,000-mile mark today! 1,002 miles to be exact.

I'm sitting at the Southside Boat Club in Koekuk, Iowa, watching the Packers get beat by the Bears in the season opener. Why do I care so much about the Packers and feel so anxious when they play? It's genetic. I know it.

I've met the nicest people here. Several different folks, old and young. At least seven or eight people have made a point of stopping by my table to ask about my trip. One guy gave me his phone number. He lives downriver about five miles. He encouraged me to stop by tomorrow for a shower and laundry. Everyone has a story about someone they've met who is paddling the river. Usually, it is a story of some hapless paddler who shouldn't have been on the river to begin with.

It's my second to last day on the river for this phase of the trip. Day 41. And it included a little bit of everything. Chi drove my bags back down to the Fort Madison municipal dock. I had a great night's sleep and was ready to paddle. The first six or seven miles were serene. The river was wide, but I decided to make the mile and a half crossing from right to left. It was relaxing and fun.

Around mile seven, I stopped at a pull-out spot in Nauvoo, Illinois, apparently a Mormon settlement. I peed and had some food. Stretched my legs. The pull out was muddy and had been destroyed by high water, and never repaired.

Back in my boat, the river turned due south and I got blasted by a non-stop 15mph wind that gusted higher, and I had that wind in my face the rest of the day. It was exhausting. Non-stop, hard paddling, and almost no progress. I went from paddling 4mph for first two hours of the day, to a crawl at maybe 1mph for the next six to seven hours.

Several times this morning I had to paddle around working tugboats and barges. There were a couple dozen barges getting moved around, so I had to paddle way out into the windy, wavey middle of the river to get around them, which was nerve-racking.

At one point, I was on the left side of the river and decided I needed to cross over to the right side since I was at the narrowest point in the river (maybe a mile wide?), and with the wind howling in my face, and the river widening to two miles up ahead, I decided it was now or never. So, I paddled at an angle to the wind and waves for the mile crossing. The wind and two-foot waves coming upriver towards me made for an anxious white-knuckle crossing.

It's hard to describe how it feels to be in a fully loaded 13-foot solo canoe, paddling alone on this massive river, heading into the wind and waves, and having to make a mile or longer crossing. Especially the first part of the crossing, when you are heading towards the middle and getting further and further from shore. You really have to concentrate on keeping your boat at an angle, so the waves don't come over the side. A single wave is enough to swamp the boat, and you and your gear, several hundred yards from shore.

It's why several boat captains and lock staff have said they think it's crazy to be on this massive river in such a tiny boat.

A couple of times I decided to just get off the river, stash my boat, and figure out a way to get a ride to Keokuk and call it quits for the day. I was done with feeling anxious and tired.

At one point, I paddled up a small stream to get to the road. I got part way up the stream and ended up on one long sand bar. So, I walked my canoe back out to the main river. Big fail!

About four miles from Keokuk, I pulled out at a boat launch and had a plan to call an Uber to haul my bags to Keokuk. I was then going to trailer my canoe and pull it those four miles along the road. That's how desperate I was to get out of the wind. It was that bad!! But once I got out and walked around a bit, I decided to just keep plugging away.

Three miles later (two hours) I was at Lock #19. A huge drop of over 32 feet!! I was able to lock right through. No other boat traffic in either direction.

So, Day 41 was full of a lot of hard and anxious paddling. A fitting end to my 40+ days of paddling. Tomorrow the wind is supposed to be out of the north, and the river narrows. So, I'm hoping to get some good miles in on my final day of this 42-day river adventure.

And I get to see Leslie tomorrow night!! Probably around 9pm.

Everyone keeps telling me how the river speeds up down towards St. Louis. So, I need to do some calculations when I get home to guess how many more days, I'll need to finish this trip. Forty-five? That's my guess.

I like the river. I like the river towns. I like how the river changes every day and sometimes a few times in one day. I like the people that I've met, and the encouragement and help they've offered. I like the hard work (once it's over for the day). I like being really focused on the weather and the wind, and on the aches and pains my body feels constantly.

I like the river!

Last Day on the River

September 6, 2019 - Day 42
Keokuk, Iowa, to Canton, Missouri – 20 miles
Total Mileage: 1,022

Today's twenty miles passed by quickly. When I arrived at Lock #20 after paddling 20 miles, I knew I was done. I'd originally thought about going six more miles to LaGrange, Missouri, or even 15 more miles all the way to Quincy. But why? I was done.

Tired.

I had a half-hour wait to get into the lock so a big barge could finish locking through. Sitting there, bobbing around in my little canoe, was when I decided to be done.

Once I paddled through, I pulled off the river at a little public campground just below the lock and dam. I walked back up to the lock to see if I could store my boat and gear overnight until Leslie and I would come back in the morning.

I ended up meeting most of the staff, and they invited me into the lock breakroom where we chatted about my trip and ate watermelon. A super nice guy, Chuck, the Lock #20 electrician, offered to drive me to Quincy, and helped me store my boat and gear behind a building and out of sight. Just one more example of the amazingly friendly and curious people we met all along the way.

Leslie met me in Quincy, Illinois, around 8pm. We went out for a beer and spent a good part of the next day exploring Hannibal, Missouri. which was really fun. We had great barbecue and beer at the Mark Twain Brewery in Hannibal, and then barbecue again for dinner in Quincy at Riverside Barbecue. The Riverside had the best barbecue sauces I'd ever had. We bought two different bottles to bring home with us.

I have mixed feelings about getting off the river after 42 days. I don't like doing things only part-way. Not finishing what I started. My only consolation is knowing that I have a plan to finish.

My plan is to go back for the first couple weeks of November, continuing south from Canton, Missouri. And depending on how cold it is, continuing for three weeks. And then coming back again for a final three weeks to continue on to the gulf. I'm not sure where I'll actually finish, or which route option I'll take to finish the river. There are several options. Mike Budd just emailed me that he finished his Mississippi River paddle last summer in New Orleans, rather than paddling on into the gulf. He stopped there "just because."

Others paddle all the way into the gulf, to Army Corps mile marker zero, miles beyond the last roads, and then catch a ride on a boat back to civilization.

And still others branch off from the congested shipping lane of the Mississippi at Fort Morgan and paddle the Atchafalaya River to the gulf. Some say the Atchafalaya is really where the Mississippi River would end if the Army Corps hadn't created the shipping route and forced the river to go a different direction. I'm leaning towards the Atchafalaya finish.

I'm anxious to get back on the river. It's part of me now.

Reflections After 42 Days on the River

It's been one week and seven days since I got off the river. One week of being back in my normal day-to-day life. It always surprises me how easy it is to go from paddling for six straight weeks, or spending a month on a big mountain, or hiking for two, three, or six months, to home chores and sitting at my office desk, or feeding my pets, running mundane errands, making and checking off incessant lists, replacing the broken garbage disposal, work teleconferences, Hoosier Action board meetings, bleaching the stained patio concrete, scrubbing and filling the hot tub in preparation for cooler evenings.

It all rushes back so quickly.

My way of easing back into regular life was to spend most of the first few days back, focusing on river stuff.

- Went through all of my paddling gear and positioned it for Phase 2
- Aired out my tent, sleeping bag, and air mattress
- Bought a more comfortable life jacket, a new boat sponge, and a modified thwart bag
- Ordered the Army Corps navigational charts for the Atchafalaya River
- Scrubbed the inside and outside of my canoe
- Unpacked, and then threw away a lot of food I'd been lugging around for weeks

- Did laundry
- Stored my canoe

That's how I got through the first few days back home: preparing to leave home and go paddling again.

I brought three waterproof notebooks on the trip and kept one in my map case; I was sure that I'd be writing in them every day while paddling, jotting down notes, short to-do lists, things I'd want to reference at the end of each day. But I never used any of them.

At home, I make list after list after list. Sometimes three or four lists in a single day.

But for 42 days on the river, I didn't have anything to make a list about.

That, in a nutshell, is what keeps drawing me to long-distance adventures—backpacking, climbing, biking, paddling.

The simplicity.

Life on these trips gets so simple that I don't even need lists.

No to-do lists.

No home chore lists.

No grocery lists.

No work deadlines or project lists.

Just living in the moment.

Lists are about the future—what I want or need to do in the future.

But paddling 1,000-plus miles on the Mississippi River over 42 days is all about focusing on the next paddle stroke, the swell or wave heading towards me, the current wind direction and speed, deciding on a camping spot, the current conversation, thoughts about lunch, aching muscles, when I had my last Advil. . . .

The monotony of tens of thousands of paddle strokes.

It's all about what's going on right now or in the very near future.

That is when my mind is the most at peace.

Sometimes . . . a lot of times . . . paddling the river was monotonous, boring, never-ending. The future moments never seemed to arrive, and river progress grinds to a halt.

But these are the moments of Zen.

Of peaceful mindlessness.

Of incredible sunsets; of sitting by a campfire drinking a glass of whiskey with Jeff; of seeing a deer swim from one side to the other; of gaining a deeper understanding of life on this river because I was actually living on this river; becoming a part of the story of this river.

Headwaters at Itasca State Park

My Wife Leslie Came to See Us Off

Too Shallow to Paddle Loaded Boats

Cousin Jeff and I Ready to Head Out

Day 2 on the Mighty Mississippi

The River Fills With Reeds by Late July

Challenging Route Finding - Day 2

Designated DNR River Campsite – Day 4

Sometimes Official Campsites are Hard to Find

Crossing Massive Lake Winnibigoshish – Day 6

Drying Out After a Storm – Day 8

Moments After Jeff Fell Backwards into the River – Day 15

Tortilla Lunch #1 – Chex Mix, Peanut Butter, and Slim Jim

Tortilla Lunch #2 – Honey, Peanut Butter, and Chex Mix

Tortilla Lunch #3 – Chili and Cheese Whiz

Portaging Around a Hydro-Dam - Day 16

Portaging a Loaded Canoe

Lock and Dam #1 in Minneapolis – Day 19

Lock and Dam #1 Pull Cord to Alert Lock Master

Our First of 29 River Locks Was Intimidating

St. Paul, MN – Day 20

Celebrating My Dad's and My Birthday in St. Paul

Passing Our First Tug and Barge in St. Paul

First of Many Paddle Wheelers We Met on the River

Foggy Morning in Wabasha, MN – Day 24

Paddling in the Rain on Day 25

Dakota, MN – Day 25

Jeff Paddling through the Fog on Day 27

Don't Ask . . . Hoochieville, WI - Day 28

Planning Tomorrow's Paddle – Day 29

It's Not Always Pretty Along the River, Clinton, IA – Day 34

Sunset Near Clinton, IA

Building a New Bridge Over the River – Day 35

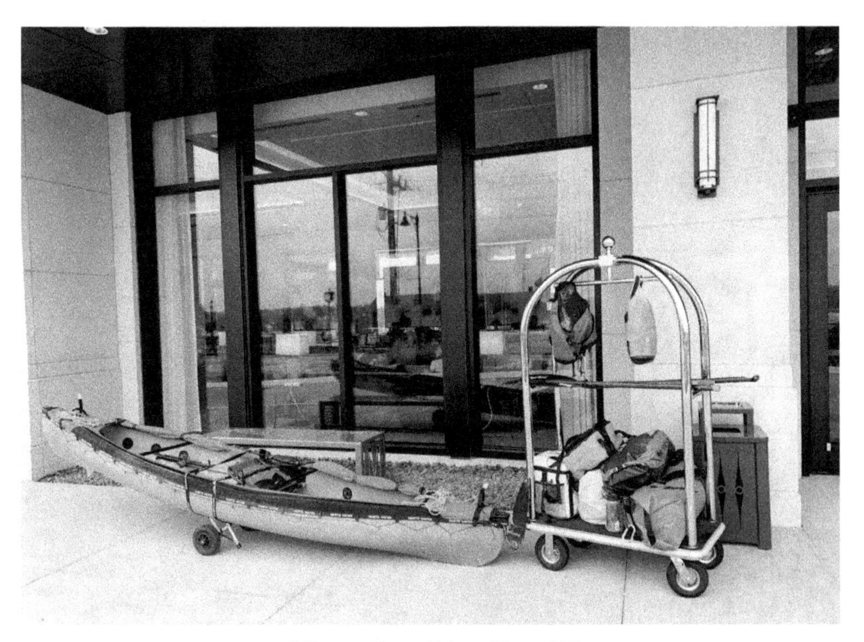

Muscatine, IA – Day 37

Camping in New Boston, IL – Day 38

Paddling Without Jeff – Day 39

Five Amish Guys Walk Into a Bar in Oquawka, IL – Day 39

Grain Elevators in Canton, MO – Day 42

PART TWO:
5 Days

November 3, 2019 - November 7, 2019
69 Miles

Back on the Big Muddy
November 3, 2019 - Day 43
Canton, Missouri (Lock #20), to Quincy, Illinois (Lock #21) – 18 miles
Total Mileage: 1,040

Back on the river. It's been 58 days since I paddled up to Lock #20 just north of Canton, Missouri, and caught a ride to Quincy with Chuck, the Lock #20 engineer. And then drove home with Leslie.

Fifty-eight days.

Two months.

I had paddled for 42 straight days. Five weeks with Jeff and then a week by myself to see how it felt to paddle and camp alone on this big river.

I would love to just finish this up. My calculations are that I have five and a half to six weeks of paddling left to get to the Gulf.

I needed to head home in September to get back to work. I was falling behind. So, I headed home for two months, all the while planning to get back to the river sometime later in 2019.

I knew it might be too cold to come back in late November and finish the river. Too cold, and too late in the year. I'm not a fan of paddling in cold weather. It makes me a little anxious. I lived for too many years in Sitka, Alaska, hearing stories of people dying of hypothermia and drowning after capsizing in the cold ocean.

And of course, I'm paddling alone, so I can't afford a mishap.

So, I bought a neoprene wetsuit top and bottom to add to the neoprene booties I already had. I also bought a new, less constricting life jacket and made myself promise that I'd wear it at all times on the river from now on.

Leslie and I drove from Bloomington to Hannibal on Friday, November 1st, with all of my canoe bags repacked, and my Wenonah Fusion strapped to the top of the car. I have pretty much the same stuff I paddled with before. A few less cooking items that I never used the first 42 days; I left behind my tarp and poles that Jeff and I only used once; I brought less food, but still probably too much; and I have some warmer clothes to match the November temperatures in central Missouri and Iowa; a much warmer sleeping bag; and, of course, my new wetsuit. Plus, fleece hat, gloves, and a neck gaiter.

We spent Saturday wandering around Hannibal. Went to a winery just out of town. Had lunch at the Mark Twain Brewery that we visited back in September mostly so we could have the "Passport to Russia" Russian Imperial Stout that we fell in love with. It's four dollars for a four-ounce glass! Twelve and a half percent alcohol. Eight bucks for the eight-ounce glasses we ordered. And $20 for a pint bottle, which I just bought four of for Leslie to bring home with her.

I spent Saturday night packing up what I would carry with me in the boat on Sunday, and going back over my maps, highlighting things to watch for, like little side channels and such.

On Sunday, we woke to a chilly, breezy morning. The forecast had the day starting in the mid-30s with a high of around 56, and sunny, but with winds 10 to 15mph out of the south, gusting to 24mph. Boooo.

I had to look up on Google the phrase "south winds" just to remind myself that it meant the wind would be coming out of the south, i.e., into my face all day long.

I'd picked an end point pick-up spot for today, just south of La-Grange at a little side channel, at what would be mile 22 for the day. We did some driving reconnaissance early this morning to check out a spot where Leslie could meet me, and found the entire area flooded.

The NOAA weather website says that this entire part of the river, from Canton down almost to St. Louis, is currently above flood stages, and there are flood warnings everywhere. For me, that means higher, faster water, and probably a harder time finding places to camp, since riverside areas may be flooded.

None of this was good news or reassuring for me.

With my pre-selected endpoint for the day inaccessible due to high water, I decided I'd end at Lock and Dam #21, just south of Quincy, Illinois.

So, we drove up to Canton at the downriver side of Lock #20 where I'd taken my boat out 58 days ago. I climbed into my skin-tight neoprene bottoms and top. They will take some getting used to, for sure. Booties, wind breaker, winter hat, light fleece gloves, neck gaiter to cut the cold wind. I felt like I was stuffed into a winter snowmobile suit.

In my boat, I only carried some food for the day, a dry bag with extra warm stuff, and another little drybag with my sunglasses, sunscreen, Vaseline, Advil, binoculars, and a knife. And I brought my portage wheels.

Leslie has been so great and so supportive of this entire trip. I told her how much I appreciated her, and then put my boat in the water, zipped up the spray deck, walked out into the ankle-deep cold water, and climbed in.

And I was off. Paddling Day #43.

Breezy and chilly. I'm glad I brought my fleece hat, gloves, and neck gaiter. At least for the first couple of hours.

For the first 30 minutes or so, I really focused on looking around, and appreciating my surroundings. I saw a bald eagle up in a tree, staring down at me. I stayed along the west (right) bank, where the wind seemed less strong. There was definitely a current. Unusual for the large pools of water between locks. But I'd heard that the lockmasters were just letting the river flow right over the dams to help ease the flooding. So, the current was faster than I'd experienced up to this point.

The current and the wind kept me alert. Sometimes I felt relaxed and contemplative, and other times anxious and hyper alert. The combination of a strong wind from the south, the strong current, and the occasional hidden underwater wing dams, really caused some weird squirrelly whirlpools and confused choppy waves.

I paddled close behind islands, and through small side channels to get out of the wind whenever I could. I was happy to be back on the water, but also a little frustrated and anxious about the wind and the unpredictable, fast water.

The sun warmed up the air, and I made great time.

It ended up taking a little less than four hours to paddle 18 miles. About 4.5mph. And that was with a 15-minute stop to stretch my legs. It's going to take a few days to get my back, butt, and arms used to this paddling business.

A couple of times I got caught up in some eddies and choppy waves that whipped my boat around suddenly. I couldn't predict it, and it scared me. I could tell I was paddling tense most of the time. It had to be the underwater wing dams, and submerged trees and rocks due to the high flood stage water levels, combined with the fast-moving water, that were causing the turbulence.

But overall, I felt good and relatively safe with my spray deck fastened down and covering the entire top of the boat, and with my wetsuit and life jacket on.

Leslie met me about a mile past Quincy, at Lock #21. After four hours of tense paddling, I was ready to be out of the wind and chop.

A good start.

Not too hard or long of a day.

I am back on the river.

I spent a couple of hours this evening coming up with scenarios and options for how far to go tomorrow. The idea of camping in the flooded areas and cold temps isn't appealing and may not even be an option. So, my options for tomorrow are as follows:

- 18 miles to Hannibal, Missouri
- 24 miles to Lock and Dam #22 and then get a ride back to Hannibal for the night
- 30 miles to the Dupont Conservation Area, that apparently has a primitive campground, which very likely

is underwater

I'll wait and see how the day goes. I have to admit that the cold weather and flooding has me rattled and anxious about paddling alone at this time of year.

Leslie leaves for home first thing in the morning.

Controlling My Fears
November 4, 2019 - Day 44
Quincy, Illinois (Lock #21), to Lock #22 – 24 miles
Total Mileage: 1,064

I was wide awake at 4am!!

My mind was racing.

Anxiety. Fear. Apprehension. Self-doubt.

What the fuck is wrong with me? These aren't feelings I allow into my head very often. In past mountain climbing expeditions, I've been able to control them much better than I am right now.

I'm worried about the cold temperatures, the icy water, and paddling alone in both. Tipping over in a whirlpool in fast moving and frigid water, with not much time to swim to shore. That's what keeps running through my mind. And then being soaked on shore, in freezing temps, as my upended canoe with all my gear floats on down the river at a pace much faster than I could catch it.

Self-doubt and fear. How much is warranted? How much is just my inability to shake it off?

I tried to rationally talk myself through each specific fear (to help me fall asleep, and at 4am), giving myself advice and positive self-talk.

Leslie was right when she said last night that once I open the door in my mind, I might only continue for a few more days rather than my planned two to three weeks. I start talking myself into the "wisdom" of the new plan. I even contemplated using the excuse of freezing temperatures and cold water to just go home with Leslie today. The option was sounding better and better the more I entertained it.

The average low temps in Hannibal, Missouri, in early November are 48 to 61 degrees. Trust me, I've checked several times. Over the next few days, the highs are predicted to be in the low 30s, and the lows in the low- to mid-teens.

Around 5am, Leslie got up to pee and I told her that I'd been awake for a while and was feeling anxious and scared. Leslie snuggled up next to me, and I was able to start to focus on other thoughts, like my tiny grandson Arlo, who was born two days ago, our upcoming summer trip to Greece and Croatia, and other distractions. I slowly started to relax my mind and fall back asleep.

Up around 6:30am. Finished packing up. Got some good coffee at 7am at Java Jive in Hannibal. Packed the car. And headed back to Lock #21 in Quincy.

The boat launch site just below the dam was muddy from the receding flood water. It took a little while to remember how I'd loaded my full boat during Phase 1 of this trip to make room for all of my stuff. Clothes pack and electronics pack in the bow; camping gear pack, food pack, and cooking stuff in the stern; map case and small miscellaneous dry bag carabineered onto the top of the spray skirt within reach in front of me; canoe wheels strapped down just behind me; spray cover zipped up tight; and my full coffee mug, lunch bag, two-liter water bottles, pee bottle, and hand bilge pump all stowed around me inside the cockpit.

Ready to go.

Leslie and I hugged.

I thanked her for supporting my crazy adventures.

Stepped into the shin-deep mud to pull my loaded canoe into the deeper water and climbed on.

It was a peaceful paddling day overall.

I paddled across to the west side of the river since the little wind there was blowing out of the west and I hoped that the bank and trees would give me some cover. I chose my angle to cross from east bank to west bank, to make sure I stayed plenty behind a tug and barges on the opposite bank that were waiting to move forward into the occupied lock.

The current was consistently flowing 3-4mph, which means my boat moved about 1.5mph with no effort from me, and closer to 5mph when I paddled.

Right away, I saw four immature bald eagles all take off from a tree as I passed by, and a big blue heron doing the same. That helped calm my nerves.

I took a couple of back channels where the water was mostly calm.

There were a few places where the currents seemed to collide with submerged wing dams and other underwater disturbances, but for the most part I felt totally comfortable, confident, and even relaxed.

I was feeling stiff and antsy to get out after 12 miles, but I pushed on to Hannibal at mile 16. In Hannibal, I pulled into a muddy public boat ramp that was closed due to nearby construction and crawled out of my boat to stretch my aching legs. The cold temps made me a little achier than I would normally be.

I walked into now very familiar Hannibal, Missouri, reserved a room at the Best Western downtown, and had a nice Reuben sandwich at Ole Planters Café.

Then back to my boat for eight more peaceful but increasingly chilly miles to Lock and Dam #22. The temp got up to close to 50 degrees, and it was overcast. But my wetsuit gets damp and clammy from perspiration, and my feet get cold from the wet booties, and it combined to make me feel chilled.

I'm wondering about this upcoming Thursday and Friday when the highs will only be in the low 30s. And the anticipated lows of 19 degrees!! That is way too cold to be sitting in a canoe. So, I'll see what I decide to do.

I'd called ahead to the lock staff and was told of a channel that I could paddle into just before the lock, so I could unload. The guy I talked with told me I could store my boat and gear on Army Corps property at the lock. I had some trouble with my new canoe trailer wheels. I couldn't figure out how to use them. How hard can it be? Two wheels, two pieces of aluminum, two straps.

I eventually figured them out after tipping my full boat over a few times. (Put on wheels, attach straps, load boat, pull it along a few feet, tips over, unload, reposition wheels and straps, re-load, pull, tips over. . . .) Day 45 on the river and I can't figure out my damn portage wheels. I feel like a rookie.

I rolled my canoe up behind the lock building, unloaded the bags, and tipped my canoe upside down, with the wheels still on, so I don't need to mess with them in the morning.

I called a taxi at 2:40pm. Not great cell service at the lock. I wasn't exactly sure when they said they would get there. The taxi finally came at 4:45pm!! After a very long, frustrating, and increasingly cold wait for a seven-mile drive back to Hannibal. Should've walked. I wasn't dressed to stand around in a sweaty wetsuit in 35-degree temps for two hours.

After a nice, long hot shower at the Best Western, I dried out my clothes in the coin-op dryer and then walked to Subway for a late meatball sandwich and barbecue chips in my room. I talked with Leslie, then Seth, then my mom and dad, all of whom said I shouldn't be paddling the Mississippi alone in November, especially with the water level so high.

I then spent my usual one to two hours examining my Army Corps river maps trying to figure out my options for how far to go and where to stay the next couple of nights.

I'm not excited about the prospect of camping along the flooded river, now that it's getting dark by 5:15pm, and getting into the teens overnight. It's 9pm right now, and if I was camping tonight, I'd have been bundled in my sleeping bag for the last four hours, and damp from perspiring in my wetsuit all day, and psyching up for the long night ahead.

There are definitely some down sides to being on this river in November. But overall, it was a great second back on the river.

Do or Do Not. There is No Try.

November 5, 2019 - Day 45
Lock #22 to Louisiana, Missouri - 17 miles
Total Mileage: 1,081

I only paddled 17 miles today. Because that's how far Louisiana, Missouri, is from Lock #22. Another ten miles would've gotten me the small town of Clarksville, Missouri, but I've scoured the internet several times and there is no place to stay indoors in or near Clarksville. And I'm not in the mood to camp if I don't have to. It's not the cold so much as the 13 hours of darkness in the tent. Twenty degrees isn't sit-by-the-campfire weather. Not into it.

I took a 7am walk to Java Time in Hannibal, where Leslie and I got our morning coffee the last couple of days. The pregnant baristas (yes, two of them), recognize me, and this morning I again showed them my latest photos of now three-day-old Arlo.

Shower. Packed my couple of bags (the rest I left with the canoe at the lock) and got in my pre-scheduled taxi at 8:15am. It was a super disgusting, filthy minivan driven by a really nice guy, who immediately lit a cigarette as soon as I got in. He made a funny joke about my Packer hat. (He's a Cowboy fan, who was talking about the Monday Night Football game last night where his Cowboys beat the Giants.)

After a brief chat about the game, the driver called a friend on his cell, and proceeded to say things like, "Ya, dat mutha fukka . . ." and "Mutha fukka don't know shit." Non-stop. It was hilarious.

Taxi etiquette is different here in rural Missouri.

I wheeled my boat outside the locked fence at the lock and dam, finished loading it, and then pulled it down to a boat launch on the downriver side of the dam. I had to get in the cold water, calf-deep, to get my boat to float off the wheels so I could unstrap them. The water was freezing, but my neoprene booties warmed up in a few minutes, and I was off and paddling by 8:45pm.

Sunny. Calm. Peaceful.

A slight breeze at my back.

The leaves are still full of fall. Lots of color.

I traded back and forth between paddling with my thin gloves on, off, and then on again. The temp started out in the low 40s but eventually got up to 50. And having the sun out made all the difference.

I'm spending too much time in my head thinking about upcoming days in the 20s and low 30s; winds out of the south; cloudy. It all makes me a little bit nervous. It's hard to stay warm in my boat. Especially my not-moving, wet feet, even when the temps are in the 40s. They still feel freezing.

I took a three-mile back channel which added a little extra distance, but it got me away from those nerve-racking wing dams. So, it was worth it. With this 2mph current, I find myself doing whatever I can to find the route with the fewest wing dams. They are marked on the Army Corps maps and show up on one or both sides of the river in various places, to channel the water where they want it to go.

Some of the submerged wing dams just cause a series of ripples in the water, but others cause small whirlpools and eddies, and cross currents that swing the bow and stern around in unpredictable jerky tugs and pulls. For brief moments I lose control of the boat and in a second can swing 90 degrees and be broadside to the current, risking rolling over.

Maybe I need to be paddling really close to the shore when I think they are coming up? With the river still above flood stages, there are also all kinds of submerged boulders, strainers, and logs right along the shore that present different problems.

Lots of herons, and some eagles today. No towns and only a couple of summer shacks on stilts.

Even with a bit of current, 17 miles takes some paddling and time. About three and a half hours into it, I rounded a bend and saw a car bridge over the river that marked the town of Louisiana.

I pulled off the river at a public boat landing that was choked with floating logs and debris. Google maps shows the Louisiana Boat Club where I'd hoped to store my boat for the night, but when I got there, it looked abandoned. Maybe damaged from the flooding.

An old African American guy named Tyrone greeted me as I pulled up to the concrete shore. I dragged my loaded boat up on the shore and then walked up to chat with Tyrone a bit.

I called the River's Edge Motel and the owner's 16-year-old son (who I found out later only had his Learner's Permit) drove down and picked up all of my bags, and I wheeled my empty canoe about a mile, mostly uphill, to the motel. It's always nice to walk a bit after sitting for so long.

Louisiana is a super cute little town with many, many large brick homes and mansions. Many were boarded up and abandoned. Some really cool looking houses. As I wandered around town later in the day, looking for a coffee shop (found none) to do some work, or a restaurant (only found one that was open) to do the same, most of the "downtown" businesses are also closed and derelict. There are actually two barbecue joints and a bakery, but all were closed on Tuesdays for some reason.

So, I had a late lunch/early dinner at a Mexican restaurant in one of those cool old red brick buildings with tin ceiling tiles. I think I was the only patron there.

I spent too many hours in the afternoon and evening scouring Google maps and my Army Corps paper maps to come up with a paddling plan that seemed reasonable, one that avoided having to camp and considered the predicted wind and drop in temperatures. My room at the River's Edge Motel was small and sort of dark and depressing. My wetsuit hanging on the coat rack to dry it out, and various pieces of clothing and gear strewn all over the place. My canoe was on its wheels, parked just outside my motel room door.

"Blustery" is the forecast for the next couple of days. There are just not many options for towns for the next 50 or 60 miles of river. And the thought of having to go 25 to 30 miles on one of those bitterly cold and windy days seems super not fun at the moment, and even dangerous.

So, I finally came up with a plan to do a series of short days: 10, 14, and 17 miles over the next three days with the option of even taking a day or two off if the weather or the situation just feels too uncomfortable or dangerous.

Once I let my mind settle on the "shorter" paddling days option, I immediately felt better.

Tomorrow (Wednesday) I'm only planning to paddle ten miles, super short, but it was either that or 24, which seemed too far given the predicted wind and cold. Then I've asked the motel owner here at the River's Edge if she could pick me up and bring me back to Louisiana for a second night. Ten miles gets me to Lock and Dam #24 (there is no Lock and Dam #23 for some reason) where I've already called ahead and have permission to leave my boat for the night. It's at a town called Clarksville.

Then Thursday, I'll get a ride back to my boat and continue on.

I've been nervous about the thought of paddling an empty boat tomorrow because the wing dams are even worse with a lighter boat, and it just dawned on me that I can bring as many bags with me as I want, to add some weight to my boat, and just leave them with my boat at the lock and dam tomorrow. Knowing I'll have more weight in my boat makes me feel more comfortable.

I am overthinking everything way too much. I'm obsessing about getting swamped in sub-freezing temps and icy water.

But I have a very reasonable plan for tomorrow!

As Ready as I Can Be

November 6, 2019 - Day 46
Louisiana, Missouri, to Lock #24 (Clarksville) - 10 miles
Total Mileage: 1,091

A short day.

A good day.

I got a little further down the river, so that's always good.

For the next 50 to 60 river miles, there are really no towns or lodging options. And the pending sub-freezing temps have me more than a little wary. Hence my low miles today, and probably tomorrow and the next day. Only 10 miles today, but at least I got out on the river and made some progress.

My trip, my rules.

The motel owner said she could pick me up in Clarksville at Lock #24, but not until around 2:30pm, so I didn't even leave my motel room until 10am. It was nice to just laze around, watch the news, and get some work done.

I got some coffee at the Shell gas station down the street and ate the rest of the coffee cake from the Hannibal bed-and-breakfast that I've been hauling with me for the last few days.

I decided to bring most of my food and gear along with me today, just to give my canoe some weight and stability. My boat was the squirreliest that first day back on the river, when I paddled it empty. So, I loaded up my boat, and wheeled it the mile back down to the city boat launch.

It was breezy, 10+mph and blowing into my face, coming out of the south. The temperature when I started was 40 degrees, so I wore my fleece ear band and light gloves.

I paddled past some industrial sites with big cranes and chutes that were loading barges with sand or gravel, or some yellow grainy looking material. Not sure what that was. Once I paddled out of Louisiana, I didn't see anything but fall-colored leaves and a couple of herons until I spotted Lock and Dam #24 off in the distance. If the river is straight, I can see a dam when it's still four or five miles away.

The wind blew steadily into my face, so I felt like I wasn't going very fast. It's been like that the last few days. But I was only paddling 10 miles, so it didn't matter.

I had the thought today that my emotions regarding the river—and the day, and the trip—jump all over the place throughout the course of a single day. I can go from feeling down and dejected, to lonely, to scared, to really at peace and relaxed, to appreciative, to bored, to confused about why I'm doing this section alone at this time of year when I could be home with Leslie, or in Michigan with Seth, Shauna, and Little Nugget.

And then at times, I have moments of clarity when I totally understand why I'm out paddling on this river. And it seems like it is the best possible thing in the world to be doing right now.

All of those thoughts in the course of just a few hours.

I made it to the lock before 1pm. Unloaded my boat along some rocky riprap, and then strapped on the trailer wheels and pulled my loaded boat to the lock office area. I ended up talking with two employees who agreed to let me store my boat and gear on the grounds of the Army Corps facility.

They had lots of questions and could not understand in the least why I was paddling the river, and why this late in the year. I told one guy that I'd already padded 1,100 miles, and he just couldn't believe it.

Some guy across the street, sitting on his porch and watching me pull my boat out of the river and get it on the trailer, yelled his support and encouragement. "God bless you." "Be safe." "That's awesome what you're doing." His encouragement really picked me up.

I also met a couple of contractors at the lock who were setting up to do some major repair work on the lock. The lock will be closed to boat traffic for three months beginning December 4th.

One guy, Charlie, offered to drive me back to Louisiana, which was great since I'd arrived an hour and a half before the lady from the hotel was supposed to come.

I've been thinking a lot the last few days about how cold my wet feet will be when the temps get down into the 20s. I'd decided I was going to start wearing warm socks inside garbage bags, stuffed into my wet shoes, starting tomorrow. But then it dawned on me that I should just buy some insulated rubber boots! Why didn't I think of that before?

Charlie brought me to a Orsheim's Farm and Garden store a couple of miles outside of Louisiana, waited while I ran in and bought an awesome pair of Muck Boots and a pair of warm winter socks, and then dropped me back off at the motel.

An unexpected river helper.

Charlie told me that he had too much respect for the Mississippi River to be out on it at this time of year, even in a big boat with a motor.

I feel like I'm being safe and smart and very alert, but I also know there are plenty of risks. Otherwise, I wouldn't be feeling the anxiety that I've been feeling.

Tonight, the temperature is supposed to drop and tomorrow the high will only be 35 degrees. So, I'm going to bundle up with a warm jacket over the top of my wetsuit top and wear long pants over my wetsuit bottoms. I have my new warm insulated rubber boots and heavy socks. And I have warm gloves, hat, and even a fleece face mask.

I checked the map and marked places where I can pull off the river if I get too cold.

As ready as I can be!

Reality Sets In

November 6, 2019 - Day 47
Lock #24 (Clarksville) to Nowhere - 0 miles
Total Mileage: 1,091

Well, the weather rules.

I got up today. The wind was blowing 20 to 25mph and the temp was 28 degrees. Cold and wintery. The wind chill was 13 degrees!

I, of course, have been obsessing about my options. Today it's cold and windy. Tomorrow even colder, and Saturday and Sunday it is supposed to warm up to high 40s. And then the bottom falls out of the thermometer and its low teens for the next week.

Way too cold to be on this flooded and fast-moving Mississippi.

I got a ride to Lock 24, felt the icy wind in my face, thought about Charlie's advice from yesterday, and decided to bag it. I could take the next two days off, and then paddle during the "warmer" weekend days, but there is no way that I'll paddle when it's in the teens and low 20s.

So, I'm just going to head home.

No reason to continue taking these risks.

Seth, Leslie, Chuck at Lock #20, my cousin Jeff, my mom, Charlie, and another guy here at Lock #24 all have been advising me that it's too dangerous to be out on the river with the combination of flood level water, fast current due to the dams all being wide open, cold temps, wind, frigid water, and paddling alone.

As Charlie told me yesterday, "I have too much respect for this river to be in a small boat, alone, at this time of year." He was a tugboat captain for over 15 years, so he knows what he's talking about.

I'll be back!!

Adding Skid Plates to My Bow and Stern

Monkey Face, My Adventure Mascot. Back on the River.

The Weather Turns Colder - Day 43

Temps around Freezing Mean Insulated Boots

Flooding Waters Near Palmyra, MO – Day 44

New London, MO – Day 45

Kinderhook, IL – Day 46

PART THREE:
43 Days

September 13, 2020 - October 26, 2020
1,080 Miles

Good For My Soul

September 13, 2020 - Day 47
Lock #24 (Clarksville) to Stag Island – 24.5 miles
Total Mileage: 1,115.5

We—my paddling brother and I—are back on the river. It's September 13th, 2020. Ten months since I was last here.

When I look back at the last few entries from November of 2019, I can't believe I was even out there paddling in 20-degree weather with wind chills in the teens. What a difference paddling two months earlier in the year, and with my paddling partner Jeff, makes on my confidence and fear factor. Today, paddling over the wing dams and the squirrelly water was a breeze.

Last November, I was terrified and tense, and constantly second-guessing myself.

No heavy boots. No tight, sweaty wetsuit. No winter hat and gloves.

God, what a difference.

Last November, being out in that weather and high, fast water alone was really stressful.

Getting back on the river today was good for my soul. I need to be out here right now. Work has been wearing on me. I've been getting short-fused and ornery.

So, it's time to clear my head, simplify my life, and just focus on mileage and weather, and sore muscles, and eating, and maps, and wind speed and direction.

I miss the simplicity.

Leslie and I got to live simply on our three-week bike trip in July, on New York's Erie Canal and through the Adirondacks, just two months ago.

So, I'm ready to get back to a simpler and more deliberate way of living.

It is mid-September, 2020. We're 50 days away from running Donald Trump's ass out of the White House; Covid-19 is still killing about 1,000 U.S. citizens each day! We're up to 193,000 deaths on Trump's watch. So, to be honest, it's nice to be taking a break from the daily news too.

Leslie and I drove to Hannibal, Missouri, Friday evening, arriving about 9pm. Jeff and Chris got here around the same time. We all walked to a local redneck bar at the end of the main street in Hannibal. It was full of people inside, with a few more outside where we sat. And not one single person wearing a mask, other than the four of us. Not one.

After a few drinks, we walked back to our B&B and had another drink in our PJs and didn't end up going to bed until after 1:30am!! Jeff and Chris seemed to still be going strong, but Leslie and I were exhausted.

Saturday morning came way too early, and we both had pounding headaches!

We spent Saturday walking around Hannibal, having lunch at the Mark Twain brewery, followed by a pizza and beer dinner, and we were in bed by 10pm.

This morning, September 13, I couldn't wait to get going. We agreed to head out of town around 8am, so good coffee at the local java place, cars packed up, and driving towards Clarksville, Missouri by 8:15am. It was about an hour's drive to Lock and Dam #24 in Clarksville where I'd decided to call it quits last November. It brought back so many memories from those few short, cold days on the water.

I have so much shit!

Loading my canoe, I was getting a little nervous that it wouldn't all fit. I have one huge dry bag full of food; a bag of electronics and cords; a clothes and hygiene bag; my big waterproof pack full of the tent, inflatable pad, sleeping bag, and collapsible chair; one small padded zipper bag for my stove, pots, eating stuff; my day bag full of stuff that I want access to during the day (sunglasses, sunscreen, lunch, raincoat, bug repellant, wallet, etc.); and then another small dry bag with maps and repair kit. Plus, two water bottles, a pee bottle, bilge pump, sponge, map case, and extra paddle. So, it's a lot to fit in my 13-and-a-half-foot canoe.

We pushed off the boat ramp just below Lock #24 at 10:20am. Leslie had left around 9:30am to start her long drive home, and Chris stayed to see us off. It's hard to describe how great it felt to be on the water, restarting this great Mississippi River adventure.

We had a slight 5mph wind at our backs. Awesome! Plus, a little bit of current. Maybe 2mph. Amazing! And it was maybe 68 to 70 degrees. Perfect paddling conditions.

We were planning to go about 21 miles today. Not too far on the first day since we weren't on the water until 10:20am. But the breeze and slight current helped us along and we ended up gong 24 and a half miles in around six hours. A great first day effort!!

It was a pretty, isolated stretch of river. We passed two little enclaves of small summer cottages all sitting up high on pilings, twenty feet or more above the ground. We paddled past the tiny village of Hamburg on the Illinois (left side) of the river. But otherwise, it was trees and river.

A big flock of pelicans just flew overhead. I'm not sure I've ever seen 40 or 50 pelicans flying in a V formation.

Oh, and around mile 20 we stirred up a whole school of carp who literally came flying out of the water all around us. Maybe 100 "flying carp" suddenly started jumping high out of the water on both sides of our boats as we paddled through. Jeff said, "flying Asian carp" are a thing the further south we go. I've never seen anything like it. Jeff said he's heard about them landing in people's boats and on their laps.

We passed several small fishing boats on the water today. But other than that, it was a calm and peaceful paddling day. After about five hours of paddling, my shoulders were getting sore along with my lower back. So, I took my first Advil of the trip.

As we started scanning the shores for a place to camp, we passed a nice clearing. And a sign that read "Primitive Campsite," but there were two guys sitting there on lawn chairs, and a third on a four-wheeler, and they didn't look very friendly. Jeff made eye contact and waved, and they didn't wave back, so we kept paddling.

We eventually found a nice little sandy beach on an island labeled "Stag Island" on the map. We have our tents set up back in the trees, and a nice campfire going on the beach. The sun just set.

We're sipping bourbon from my new set of plastic indestructible bourbon tumblers. All is good.

Unfortunately, it's only 7:15pm and I'm ready for bed. So, we'll see how long I can make it.

Crazy Mike and His Unicycle

September 14, 2020 - Day 48
Stag Island to the Yacht Club of St. Louis – 26 miles
Total Mileage: 1,141.5

Today seemed a little more like work.

We paddled 26 miles, which is a great Day #2. A few more miles than yesterday. There was a wind in our faces for a good part of the day. Not too bad. Maybe five to eight miles per hour, but enough to make it feel like the effort put into paddling wasn't quite equaling the speed or distance attained.

I also really need to stop a few times a day and stretch my legs. Jeff could sit in his kayak all day long and never get out. Not me. I also really like taking a lunch stop somewhere for an hour, just to take a break from paddling. We really didn't take any breaks today, so by 3pm I was done. I was in my boat by 8am, so seven hours later, my back was sore, and my legs were aching.

Let's go back to the start of the day.

Since I was in my tent by 8:15pm last night, I got plenty of sleep. We spent the evening by a beach fire, and finally got driven into our tents by the mosquitoes.

I tried to start my book, *Life on the Mississippi* by Mark Twain, but just couldn't keep my eyes open. I slept until around 5:30am and then just laid in bed dozing until it was light enough to get up around 6:30am. Hot instant coffee and a banana and some coffee cake from the B&B in Hannibal.

It takes Jeff a while to get his boat packed, so I sat and reviewed the maps for a while, and then just pushed my boat off and drifted downriver until Jeff caught up about a half-hour later. It was a cool, cloudless morning, but it didn't take long for it to heat up.

There are virtually no houses along this stretch of river today. Just mile after mile of trees. A big 15-barge tow pushed past us, heading upriver. We got to Lock and Dam #25 at about mile seven for the day. I pulled the cord that hangs a few hundred yards before the massive lock gates, three or four times, which usually sets off a very loud buzzer or bell. No one at the lock responded through the small speaker that hangs above the cord, and we couldn't see any staff walking around from our water level boat view. I finally looked up the phone number for the lock and called. They had no idea that we were sitting there waiting to lock through. I guess the buzzer wasn't working.

About twenty minutes later, the massive gates opened, and we paddle in. No one tossed us a rope to hang on to, and we never saw a single person. Sort of weird.

I remembered the first lock we went through in Minnesota last year. We were both so nervous. These massive chambers with millions of gallons of water flooding in or out while we sit in our tiny insignificant boats.

Just after paddling out of the lock, we stopped for a few minutes to get out and stretch our legs. And then again, around mile 14, we pulled off so that Jeff could poop.

The wind started picking up and blew into our faces. It made the 80-degree air feel cooler but slowed us down a bit. There was less of a current today too, maybe only flowing about 1mph.

The map showed that we would paddle past 10 to 12 marinas or harbors today. So, I was excited to either stop somewhere for lunch at a marina café, or at least stop for a cold pop. But every "marina" we paddled up to, ended up just being a few boats tied up in front of a few rundown houses on stilts. Not the marinas I was imagining. So, we just kept on paddling, and hoping the next one would be more inviting. We were originally heading to a riverside restaurant around mile 28 so we could have a nice meal after stopping for the day. But Jeff checked the internet, and it wasn't open on Mondays.

At mile 26, we got to the Yacht Club of St. Louis. Lots of big, beautiful boats and a nice grassy lawn. I found a phone number and called the yacht club office. The guy said that we could set up our tents in the grassy area. There was a nice gazebo and lots of outdoor tables and chairs. And no one was around. So, we immediately spread our shit all over several of the tables and chairs to air things out and just make things easier to get to. Jeff did some repacking and reorganizing. I charged my phone using a solar charger. It was so nice to sit under the gazebo and out of the sun. We roasted in the sun all day today. I definitely used my 100SPF sunblock.

Dinner was Spanish rice, a bag of pre-cooked baked beans that just needed to be heated, a can of jalapeños, and a can of Spam, all mixed together. It was actually pretty good. We drank our remaining two beers, and now are having a tumbler of Redemption rye whiskey while I write and Jeff blogs.

Today was harder work than yesterday. But I just need to get my back, butt, and shoulders back into paddling shape. We went plenty far and have a great place to camp, so all in all a great day.

Around dinner time, a guy on an oversized (tall) unicycle, Mike, pedaled over. He was a little odd, and proceeded to tell us his entire life story, part of which included assuring us that he was stoned out of his mind. Which explained part of his weirdness. He's a local

guy. A self-described "river rat" who grew up around here.
Mike said, "When we were kids, like eight years old, we'd play on the sandy beaches and drive our boats up and down the river. We thought this was the Caribbean. We didn't know any better."

Crazy Mike on his extra tall unicycle. He was pretty enamored by our trip, but he couldn't stay focused on any particular topic. Jeff and I were both relieved when Mike pedaled away.

Tomorrow we are heading to Alton, Illinois, to stay in a town!

Davie the Pirate and Five Best Things
September 15, 2020 - Day 49
Yacht Club of St. Louis to Alton, Illinois – 24 miles
Total Mileage: 1,165.5

Let's start with the good stuff.

Around mile 12.5, we stopped at a marina in the town of Porte du Sioux. I was ready for a break. We'd originally planned to stop in the town of Grafton, around mile seven, but it was out of the way, and on the Illinois (opposite) side of the river, so we paddled right past it.

I pushed off this morning at 8:15am. Jeff seemed like he was almost ready, so I just floated with the slow current and read my book, waiting for him to catch up. And, for reasons that aren't worth expounding on, Jeff didn't take off until around 9:15am. So, I'd already been slowly drifting for over an hour before he caught up to me. Consequently, my ass was especially sore by the time we stopped for lunch around 12:30pm.

As we paddled into a marina in Porte du Sioux, and past scores of sailboats and yachts tied up in their slips, I asked an older guy standing on his boat if there was any place in town, or at the marina, to get a cold drink and something to eat. He told me about a café in town, and offered to run home, and bring us back some cold lemonade. He also offered to drive us to town. Since we were both more interested in just stretching our legs a bit, we paddled on to the transient dock at the very end of the last slip.

A few more boats down, Davie, an African American man with dreadlocks tinged with strands of grey, struck up a conversation with Jeff as we paddled past. Davie offered us cold beer and before I could say "hell yes" Jeff said we still had miles to paddle and probably shouldn't.

What?

So instead, Davie brought us out four ice cold bottles of water. Really nice, especially since it was another hot and muggy day. Davie was interested in our trip and asked lots of questions. Super friendly.

Davie had pirate flags hanging all over his boat. I asked him if he knew any pirate jokes and told him that my wife has a million of them.

Davie the pirate was the highlight of my day. Jeff called him a River Angel and exchanged contact information. I called him a river pirate!

We tied off at the transient dock and ate our lunch at a picnic table sitting just uphill from the marina. I had a peanut butter and jelly bagel, two of Seth's deer jerky sticks, and Jeff shared some of his dill pickles.

We stopped to visit with Davie on our way back out of the harbor and thanked him again.

The second highlight of the day is music. As the day wore on, and our final 12 miles seemed to take forever, I put in my wireless ear

buds, and cranked U2, and then some powwow songs.
They got me through. Especially the pounding, rhythmic, heartbeat of the powwow drums. I felt powerful. It was weird. I found myself thinking, "I feel powerful."

Okay, the third best part of the day is happening right now! We are in Alton, Illinois, and we happened upon an Irish pub with outdoor seating right in downtown Alton. Complete with a five-piece traditional Irish pub band. Tin whistle, guitar, some kind of tiny bongos, a flute, and an Irish hand drum. Perfect.

I'm eating an amazing lamb-meat burger and Jeff got some kind of Irish fish dish that was literally shipped over from Ireland. What a great way to end the day. Jeff just said, "This is why we're paddling the Mississippi." I couldn't agree more.

Okay, I just remembered my fourth great part of today.

Jill!

Jill is the staff person we met at the Alton Marina. I paddled up after a long day on the water, and Jeff was still 45 minutes behind me. I walked up to the marina office and met Jill. She was an older-than-middle-aged marina worker who walked with a limp and turned out to be the nicest person on the planet.

After welcoming me to the marina with an ice-cold bottle of lemonade, Jill offered us a spot to store our gear and boats, and helped me find the Cracker Factory, a nearby little boutique hotel. She also offered us the use of the marina showers and laundry, and was just a funny, pleasant, helpful, friendly human being. She made our entry into Alton so much easier and smoother after a long day of paddling.

Okay, a fifth thing just popped into my head. Mike is the owner of the Cracker Factory. Jill suggested that I give Mike a call, and when I did to inquire about a room and told him about our trip, he was more excited about it then I was. He kept saying, "Right on. That's so awesome."

Mike, who we never ended up meeting face-to-face, helped make today a much better day. As we were making our way to the Cracker Factory with the few things we'd need for the night, including our dirty laundry, I called Mike to make sure we were walking in the right direction. I told him we'd just climbed over a street barrier that said, "Do Not Enter" blocking the entrance to the foot bridge over the highway that we needed to take, to get from the marina to the Cracker Factory, Mike said, "Right on. That's so awesome."

So, that was Day #49 on the river. I don't want to bore myself with details of when we woke up, or how far we paddled. Writing about five great things is more than enough.

The People Make the Trip

September 16, 2020 - Day 50
Alton, Illinois, to Mosenthein Island (6 miles north of the St. Louis Arch) - 17½ miles
Total Mileage: 1,183

Paddling this river is as much about the people we meet as anything else.

- **Mary:** A 72-year-old lady who was the night watchwoman at the Alton Marina. When we arrived this morning at 7:15am to retrieve our gear and boats, she seemed so happy to see us. She helped carry our bags and gear to the boats, filled all our water containers, gave us each an ice-cold bottle of water, kept admonishing us to be careful on the big river, and all the while telling us about her five grandchildren and eight great-grandchildren. Mary just couldn't believe that we were paddling all the way to

New Orleans.

- **Luke:** We had only paddled a mile and a half downriver from the Alton Marina when we reached the Mel Price Lock and Dam. Jeff called as we got close, and the lockmaster said that it would be a minimum two-hour wait while he locked through two huge southbound barges. So, we decided to portage around the lock, more for something to do than anything. I don't do well just sitting.

  What a mistake!

  We unloaded all of our shit and had to haul our multiple loads and boats up a steep, brushy hill to a paved walking path. At least four trips. Jeff only unloaded half of his boat, so I helped him carry his still half-full boat from the river up the hill. Ten steps. Take a break. Ten steps. Take a break. Then we mounted our boats on portage wheels, reloaded them, and towed these heavy monsters about a mile and a half to a spot where we had to clamber down lots of loose rock, and then re-load on a shore where we sunk down into 12 to 16 inches of very soft mud, making several more trips back and forth. Exhausting. Time consuming. Pain in the ass.

  At the spot where we put our boats back in the water, below the dam, sat a young guy wearing a U.S. Marines t-shirt who was doing some fishing. As soon as Luke spotted us, he offered to help. I said, "Oh, no. We can get it." But after the first load, I took Luke up on his offer. He said, "I'm happy to help you guys." He was totally into the details of our trip. At one point, Luke said, "You guys are so inspirational." Standing in soft mud up to my calves, I wasn't feeling very inspirational. Luke said, "My dad has always talked about going on a backpacking trip. I can't wait to tell him about you guys."

Luke was awesome, positive, helpful, and encouraging. He also caught a huge fish while we finished loading our boats. He said that it was his "P.B. Carp." His personal best.

- **Jeffrey:** Jeff got some info from a local paddling guide, that rather than paddle the eight- or nine-mile-long narrow, barge-clogged, man-made canal that led to our final Lock #27, we could take out just before some well-known rapids, portage around, and get right back on the river. So, that's what we did. It involved another unload, short portage, and reload, but it was only a few hundred yards, and the reloading spot below the rapids was nice and sandy.

  While I was carrying my first load across a parking lot, an unusually tall African American guy, an old man who was getting in his car, stopped and asked if he could help haul our stuff. He said, "We can throw that canoe right on top of my car." I didn't take him up on it, but so nice!

  On my second trip across the parking lot a Hispanic guy sitting in a beat-up truck asked where we were going and when I told him about our trip, he said, "You got big balls. Bigger than mine. I would never do that."

  Downriver, where we set our boats down on the sandy shore to reload, there was an old African American guy and his wife fishing for catfish and sturgeon. His name was Jeffrey. Jeffrey showed me his string of five little sturgeon and even posed for a photo with a catfish that he'd just caught. After telling him about our trip, he kept referring to our truck up in the parking lot, and asked, "How you gonna get your truck down the river?" I tried to explain a couple of time that we'd paddled here, and were going to paddle away, but he just didn't seem to understand. His wife was super nice too. Jeffrey told her, "These guys are paddling to Minnesota!" He just couldn't wrap his mind around what we were doing. Most people can't. His wife kept telling us to stay safe and be careful. Delightful people.

- **Mike:** Okay, I never met Mike. He's a river guide that lives in the area. Jeff had been texting him and called him twice to find out any river info Mike would offer. This was especially important since we were not paddling in the man-made barge channel, and there were some rapids and channels that we wanted advice on. Mike told Jeff, that the best place to camp on this stretch of the river was two miles downriver from where we'd just portaged past Jeffrey and his wife. So, we headed to the spot and arrived around 2pm. The spot consisted of a huge expansive sandy area in the middle of the river, with no shade and no trees, and it was sunny and HOT. To stop here, would require us to haul all of our gear and boats at least 200 to 300 yards to get high enough above the water line to feel comfortable that our stuff wouldn't be washed away with a rise in river level, and then half of our gear another several hundred yards to get to the nearest tree for some shade to camp.

I was thinking no way am I hauling all of our stuff and our boats one more time today to what looked like a shitty camping spot. And the thought of just sitting in the sun for the next five hours, after being in the sun all day already, made me feel sick. We debated back and forth. Jeff kept saying, "Mike says this is the best spot." I thought it was a horrible spot. Mike. I finally talked Jeff into calling Mike back to ask about another possible "great camping spot" that had some shade further down the river. He de-

> scribed another site about a mile downriver, which me gladly and Jeff begrudgingly paddle to, but never found. Our paddle included an exciting set of fairly shallow rapids that we both had some anxiety about shooting with our heavily loaded boats, but it ended up being fine. We ended up finding a really nice beach spot at the far south end of an island that had some nice shade. And we only had to haul our stuff about 100 feet.

We just finished a noodle and tomato sauce dinner that Jeff brought, with some caramel chocolate that I brought, and our nightly tumblers of whiskey. From our beach site we can see the St. Louis skyline and the lit-up arch, about six miles away. It's a perfect spot. I am so glad that we passed up Mike's suggestion.

We only paddled 17 miles today, but the two portages were lots of work. We stopped at 4pm, set up camp, and then I went for a swim with my clothes on to cool off and rinse out my sweaty clothes. Tomorrow we will get up and paddle right through downtown St. Louis. Should be fun!!

Beneath the Arch, The Scariest Day So Far

September 17, 2020 - Day 51
Mosenthein Island to Crystal City, Missouri - 36 miles
Total Mileage: 1,219

Jeff was up at 5:30am today to get all packed up. No matter where we are, it takes him a few hours to get his gear packed and stowed and be ready to paddle. It just does. I don't get it. He was up an hour before me. I rolled out of bed at 6:30am. I'd been hearing scraping noises outside my tent before the sun came up, but I thought it was an animal. It was just Jeff bumping around in the pitch dark.

Today was a big day. Our first day with some current. We camped on a great sandy beach that was only six miles north of St. Louis, so after dark last night, we could see the city lights off in the distance. It was beautiful.

Cold cereal and coffee and a breakfast bar.

Our tent flies were soaked with dew, so I just strapped it onto the spray deck on the back of my boat. Sand sticking to everything that was wet. It's the only downside of camping on sand. It sticks to everything and gets into everywhere.

The current swept us quickly into and through the city. It was really cool to float past the arch. We took lots of photos but didn't stop. The river current probably added 2mph to our 3 to 3.5mph paddling speed.

From the arch south, for the next 10 miles, the barge traffic was heavy! The river narrowed, and there were scores of tugs and hundreds of barges on both sides of the river. Hundreds and hundreds, either tied off into large rafts, in various stages of being loaded or unloaded. And the fast-moving water was squirrelly with waves coming from all directions. Waves of two and a half to three feet were coming off the tugs as they passed that then crashed against the sides of the parked barges, pushing the water back in the opposite direction, and all perpendicular to the flow of the river.

I was really tense and a little scared for those entire 10 miles with these crazy big waves knocking my little canoe around in directions I couldn't predict. I felt like my paddling was constantly trying to react to my bow or stern getting suddenly jerked one way or another, all the while trying to keep waves from coming over the side. And it would've been a dangerous place to dump with all the boat traffic around. We were both fighting to stay upright, and were never really close enough to yell for help if one of us would've needed it.

Once the water started calming down, I said to Jeff, "Very little of that was fun for me."

But we were moving. Faster than any of our previous 50 days.

As we paddled south of the arch, there were lots of homeless camps, made up of tarps and wooden shelters, lots of industry, and lots of tugs and barges. We are always trying to decide which side of a stationary barge to paddle around and guessing which way an oncoming barge was going to turn.

But it was fun to be going faster than 5mph for a good part of the day.

Around mile 22, I had to stop and get out of my boat to stretch. It was only like 12:30pm. I snacked on and off when the waves weren't too big, but really needed to stretch my legs and walk around a few times during our paddling day. But once Jeff gets in his boat, he doesn't really like getting out of his boat. His knees have been bothering him, so he'd rather just sit in his kayak and take a break while floating around.

At mile 28, we pulled off at a little boat landing by the small town of Kimmswick. We were just going to walk the half-mile into town to get a cold pop. But it ended up being a cute little tourist town. So, we decided to stop into the Blue Owl restaurant for lunch.

The owner, Mary, was super nice and very impressed with our trip. She told us all about the town and its renaissance over the past 20 or 30 years. Mary's Blue Owl has been featured in *Oprah Magazine*, *Good Morning America*, etc.

Jeff ordered the daily special of chicken-fried steak and mashed potatoes, and I got a Reuben sandwich. It was a great little unexpected stop.

The final eight miles to our destination at a little marina in Crystal

City was hard for me to get into. We pulled in around 4:30pm, and I was ready to be done. Thirty-six miles. Our biggest mileage of the entire trip so far. We pulled our boats out of the water, and almost immediately an old hard-of-hearing guy, Tom, drove over to talk to us. He spotted my Wenonah canoe and said, "I bet you have a story to go with that canoe." What a great line.

Then two younger guys pulled up in a jacked-up pickup and said, "Hey, are you the guys who we heard were just at the Blue Owl earlier today?" What? Jeff had apparently written a post on Mary's website, and these two guys in Crystal City just happened to see it, and then saw us pulling our boats out of the water eight miles downriver. They offered us two ice-cold Miller beers in metal cans that were shaped like a bottle of beer with corn cobs painted on the sides of the can. They were pretty amazed with our trip and thought it was so cool that they'd just seen Jeff's post from the Blue Owl.

Tom offered to drive us into Crystal City, and for the last hour and a half, Jeff and I have been at the Crystal Tavern Bar and Grill. We just finished wolfing down a couple of drinks and a shrimp, potato, corn boil platter all swimming in butter.

I had a nice long phone call with Leslie, and we looked over the map for tomorrow. Now we have a two-mile walk back to our tents and boats where we're camped.

Jeff is talking about how we could have easily gone 40 miles today. I agree that we could have done it, but not "easily." He seems to be able to just sit in his boat all day long. It's not as easy for me. I need to get up and move around.

Our options for tomorrow if we want to stop at a town are going 27½ miles or 40 miles. I'm pushing for 27½! We'll see what the river gives us. It's been five days and 128 miles since we restarted our journey. I'm getting into the river groove. Might need a shower and laundry again in the next day or two.

Some People Just Don't Get It

September 18, 2020 - Day 52
Crystal City, Missouri, to Kaskaskia River - 32 miles
Total Mileage: 1,251

Fuck.

Leslie just texted me that RBG passed away. I can't believe it. I can't even imagine what bullshit Mitch McConnell and Trump will now pull to get Trump's third supreme court justice installed. Oh my god.

We are camped on a sandy bank along the river, about six miles before the little town of Chester. Our Plan A was to stop in St. Genevieve and get some dinner and possibly even a motel, and Plan B was to paddle on to Chester. I guess we opted for Plan C.

(Pause . . . several flocks of pelicans have flown over our heads since we set up camp. I'm not sure I've ever seen a flock of pelicans. They fly in a "V" shape formation like they've totally got their shit together.)

We have a nice beach fire burning. Our bellies are full of a baked bean and risotto and spam dinner. Eating chocolate chip cookies. It's a calm night and getting chillier by the minute. It might be cold in the ol' sleeping bag tonight.

I have to pause and remind myself that I'm paddling the Mississippi River! Camping on a sandy shore. Sitting next to a driftwood beach fire. We paddled 32 miles today. Nothing to sneeze at. Overall, a great day on the river.

So, why didn't we take the preferrable Plan A and spend the night in St. Genevieve? Good question. I was really excited about stopping there after paddling 27 miles. We'd had a headwind for most of the day, so I felt good about our effort to get 27 miles.

The "marina" that we were heading for looked to be back about a quarter mile up a small channel off the river. We'd planned to paddle to the marina, camp, and hike into town for a meal, or at least a grocery store. We paddled into the channel and within 100 feet it narrowed from 20 feet wide to about six feet wide. Mud had caved in the banks of the channel, probably from the big flood last year. Jeff went first, and we paddled another 200 yards as the channel continued to narrow down to only three feet wide (my canoe is about two and a half feet wide).

We kept creeping forward, hoping the channel would eventually widen. But Jeff's kayak eventually got stuck in the mud. We could see a parking area and partially destroyed marina building a few hundred yards ahead. Jeff couldn't really get out of his kayak, so I got out of my canoe, and immediately sunk up to my knees in very soft mud. It was all I could do to keep my shoes from getting sucked off my feet in the mud as I tried to move up the bank.

We eventually conceded that we couldn't go any further and started pulling our boats backwards as we retreated through the mud. Forty-five minutes later, when we'd gotten back to the river, we were both covered in mud, pretty much up to our waists.

Very frustrating, and very hilarious.

So, no St. Genevieve. No camping at a marina. No restaurant. Not even a pop machine.

At that point, we were both ready just to find a place to camp. Paddling another 13 miles to the town of Chester was out of the question. We eventually stopped at a nice sandy beach, just before a small river entered the Mississippi, and after paddling a total of 32 miles for the day.

A good solid effort resulting in a great camp spot. We spent quite a bit of time washing the mud off our of legs and our boats.

We have 66 miles to Cape Girardeau, a big town where we'll get a hotel, showers, and laundry for sure. So, that means 33 miles per day for the next two days. There is a tiny village called Grand Tower, 38 miles downriver, that we'll shoot for tomorrow. There is a gas station and a convenience store there, which sounds pretty amazing right about now.

We are low on drinkable water, and couldn't charge anything today, so phone and laptop batteries are getting low. We need fresh water and some electricity. So, if it's not too windy, we'll go for 38 miles tomorrow.

This river life is so unique.

Most people don't get it.

Paddling. Mud. Wind. Aching shoulders. Barges. Pelicans. Whiskey. Camaraderie. Paddling. And more paddling.

I'm not sure I even like paddling.

At least not seven to eight hours a day. Every day. Day after day.

But it's all of it. Everything combined. The mud and the pelicans, and my sunburned legs, and drinking whiskey by the fire, and doing all of this with my cousin Jeff. All of it combined. This is what I dream about when I'm sitting in my office.

Most people don't get it.

A Grand (Tower) Welcome
September 19, 2020 - Day 53
Kaskaskia River to Grand Tower - 38 miles
Total Mileage: 1,289

It's 6:30pm. Phew. I'm exhausted!!

We paddled 38.3 miles. A new record for us. We were on the water for eight and a half hours. That's a long damn time. Our goal was to get to Grand Tower, and we made it. Grand Tower is a tiny village, just a few blocks in each direction. We walked through most of it and saw the fire department, city hall, post office, and a gas station with convenience store. That's it. Not a lot.

There used to be a campground, but last year's flood wiped it out. So, when we paddled up to the boat ramp, there were two cars parked on the ramp just a few feet above the water line, with three locals fishing. The cars were so close to the water and taking up the whole boat ramp, so we couldn't pull our boats out of the water. The fisher-people didn't seem to take any notice of us and made no effort to move their cars. We ended up hauling them out on a very muddy shoreline a hundred feet upriver. Jeff actually asked the people, "Do you think we could squeeze our boats between your cars?" We clearly needed to land our boats and unload. But nope. They didn't budge.

The town is protected by a 30-foot-high levee wall made of earth. And on top of the levee is a little covered picnic shelter with a total of one picnic table. So, we hauled all of our stuff up to the shelter and set up our tents on top of the grassy levee in plain view of the houses across the street. We figured that if someone didn't want us camping there, they could come up and tell us to move on.

We had lots of space to spread out our stuff on grass and draped over the shelter railings. It was nice to not be camping on sand for the night. And it was nice to have a place to sit that was not our camp chairs.

After we set up our tents, we walked to the gas station, bought two Hunt Brothers pizzas and some soda on ice. Jeff also bought a six-pack of Miller Lite. Yuck!

The paddling was a little monotonous. I tried to set lots of small goals: paddle to the next red navigation buoy, to the next green buoy, to the next red buoy (see what I mean?). It's actually really beautiful country. Not a soul. All trees. We passed the little town of Chester around mile eight. I got out and filled two one-gallon water jugs in case we ended up camping tonight. Neither of us like being low on drinking water. Jeff stayed in his boat, and I got out at a concrete boat ramp. I walked to the nearest house, and it had a sign that said, "Welcome to the Hospitality House," with a Bible verse beneath.

I put on my COVID mask and knocked on the door, and Amina answered. She's the caretaker of the Hospitality House. It is lodging for families who are visiting inmates at the nearby penitentiary. Amina said that she's had no guests since COVID hit and welcomed me in to fill up my water jugs. I'm sure we could have spent the night if it was later in the day. Amina was super nice and welcoming.

The boat traffic today was surprisingly heavy, given that we aren't near any big city. We were passed by at least 10 tug tows, the biggest ones we've seen yet, now that there are no more locks. We saw a 28-barge tow (seven barges long and four barges wide); and then a 35-barge tow (seven barges long and five barges wide). Jeff estimated the tow was about one-half mile long!

We did have one hairy situation. For some dumb reason, we decided to paddle across the river from the right side to the left. There was a huge tow downriver, and heading upriver towards us, but it had to be over a mile away. But as we started cutting across the current to paddle to the other side, we realized how wide the river was at the point where we were crossing and how quickly the massive tow was approaching us.

We both paddled as hard as we could for about 10 minutes. Super nerve-racking. I was nervous that we wouldn't make it, and Jeff was behind me. We vowed never, ever to do that again. Dumb.

Last night at our beach camp, we saw several flocks of pelicans fly overhead at least a dozen times. Always flying in a V-formation. Well today, I rounded a bend and there had to be several thousand pelicans sitting on a beach. I paddled as close as I could, and when they finally noticed me . . . holy shit! There were thousands of pelicans taking flight all around me. It was really cool.

We finally got to Grand Tower around 4:45pm. The town is named after a big rock that sticks out of the river and has trees growing on it.

Long day. Long paddle. I was in the midst of bemoaning the hassle of having to partially unload our boats, and then wheel them several hundred yards uphill to the top of the earth levee, when a local young man, Charles, stopped and asked if I needed any help.

Of course!!

Charles was a super nice guy. And as we were setting up for the night, I noticed the guy across the street from where we were setting our tents up on the levee, just staring at us. So, since we were preparing to camp, just across the street from his front yard, I walked right over and struck up a conversation. I figured that if he didn't mind us setting up on top of the levee at a public covered picnic area, then nobody would. Plus, he could keep an eye on our stuff when we walked to the local gas station for pizza. Kyle (the guy across the street) was a good-ol' boy who just wanted to talk about the big fish he's caught in this river. Super friendly.

We walked about three quarters of a mile to the gas station to get our pizza, pop, and beer, and Charles was there, happy to see us again. He said that he was worried about us leaving our stuff unattended, back at the levee, so he offered us a ride back once we stocked up on food. A super friendly 17-year-old who had just joined the Illinois National Guard after graduating from high school two months ago. We thanked Charles a million times and let him know that he totally made our day.

I now have a belly full of pizza and Mellow Yellow pop. I'm about done journaling and sent emails to a couple of Airbnbs in Cape Girardeau. So, hopefully we'll find a place to stay there tomorrow, and more importantly, a place to safely store our boats for the night since we'll be in a bigger town.

It was a long paddling day, but a good day full of helpful people: Amina, Charles, and Kyle.

The Mayor of Biscuits and Gravy

September 20, 2020 - Day 54
Grand Tower to Cape Girardeau - 28 miles
Total Mileage: 1,317

We're sitting at the Minglewood Brew Pub. It seems like we were never roasting in the sun, paddling endlessly in what felt like molasses listening to the Packer game on the radio and waiting and waiting for Jeff to get packed up (he shoved off at 9am). Seems like none of it ever happened. I'm drinking a good IPA, eating a Cuban sandwich with some mac and cheese on the side, and a Caesar salad. Enjoying the moment.

Jeff had a crabby day today. Which was nice for a change. He's usually so annoyingly positive. Overly positive. But today, he was CRABBY. I was the one saying, "Come on, Jeff. You got this." He said that nothing was going right this morning. Since he was still getting his act together at 8:30am, I just started paddling. We finally hooked up around mile seven. And he was still crabby!

I don't remember if I wrote about this yesterday, but the Mayor of Grand Tower, Randy, stopped by after dark to check us out. We were sitting at the picnic table under a covered gazebo with our headlamps on, reading and writing and going over maps.

Randy pulled up in his pickup, got out, and said, "What exactly is going on here?"

He ended up staying and talking for more than an hour . . . about his community, the loss of commerce and jobs, and about some of the colorful locals. Super nice guy. We couldn't get rid of him. He told us that the guy living across the street was crazy, and that his wife was even crazier.

We had just boiled our water for coffee this morning and were packing up around 7am when Mayor Randy pulled up with his arms full of coffee and "gravy" (biscuits and gravy) from the local gas station. Wow. I had never actually eaten biscuits and gravy. Ever. I always thought it was kind of gross. But I ate every bite. It was delicious.

It was cold last night. Forty-five degrees when we got up this morning. I was cold all night long and woke up at some point and put on a long sleeve shirt, dry socks, and my fleece ear band. Everything I have is damp, including my clothes, sleeping bag, and my sticky skin. So, I was just damp and clammy, and uncomfortably cold. Not a good night's sleep for sure. And starting at 4am, the across-the-street neighbor's crazy wife started yelling. Yelling!

This morning Randy said, "I told you they were both crazy."

Also, as we packed up, Fred Houston stopped by. Another curious old-timer that spotted us. We're big news in Grand Tower. Fred told us all kinds of local stories. Super friendly. He's met lots of Mississippi paddlers over the years. He left to go make breakfast for his wife of 52 years. And afterwards, Fred came back to chat while Jeff was loading his boat. Jeff later claimed that he couldn't launch because of Fred.

As I was wheeling my boat from our camping spot to the river this morning (Fred still talking with Jeff at the gazebo), another older gentleman in another pickup, stopped and asked if he could get us anything at the gas station before we headed out.

I am 100 percent sure that Grand Tower is a super redneck little town. But these old rednecks were super nice, and very willing to lend a hand. A good lesson for me to relearn.

The 28 miles today were non-descript. Trees. Water. A few big tows passed by in both directions, which keeps us on our toes. We were done paddling by 3pm. So, a little slower pace than the last few days.

I'd contacted an Airbnb in Port Girardeau, and Rocky met us at the city boat ramp with his pickup. He loaded up our gear and boats and drove us to the cutest little two-room brick house a few blocks from downtown with a laundry next door. We were able to spread out our wet tents and my damp sleeping bag. Shower. Cold pop and water in the fridge.

Rocky was super nice and offered to help haul our stuff back to the river tomorrow morning.

The Airbnb was a perfect place to stay. It's only 5:45pm and we've finished dinner and my second IPA. I still have work to do on my computer, so can't drink too much more, but I could totally fall asleep right now.

Good day.

I'm sore, so I know I did something.

Only 50 miles to Cairo, and the official start of the lower half of the Mississippi.

Rocky the Vet

September 21, 2020 - Day 55
Cape Girardeau to River Mile 19 (Upper Mississippi) - 34 miles
Total Mileage: 1,351

We are camped at a total shit-hole camping spot along the river. We were shooting for a spot on the map that showed some kind of boat ramp, but we must've missed it. We'd already paddled around 34 miles, mostly into the wind, and had been on the water for eight hours when we finally stopped.

So, we pulled over at a spot that looked like a sandy landing, and it ended up being super muddy. We were too tired to get back in the boats and keep paddling, so we wadded through the mud to unload our boats. The shoreline was covered in river trash. Within a few minutes, I found a whiffle bat and ball, several plastic 50-gallon drums, and all sorts of other stuff. I also found a cool big turtle shell, and some neat hawk feathers.

It was a depressing end to a long paddle, but now that camp is set up and the sun is setting, and we have a nice beach wood fire burning, everything seems good. Settled. We paddled as far as we wanted to go and are all tucked in for the night.

One day at a time.

I had thought that we "only" had 31 miles tomorrow to get to a state park. But I just recalculated, and its 35 miles. So . . . another long day. But, if the park is open, with picnic tables and bathrooms, it'll totally be worth it. (I just re-read that last sentence and it sounds pretty pathetic.) Something to paddle towards. It's hard for me to stay motivated all day long when there isn't an interesting destination to get to.

We saw zero people or small boats today after we pulled out of Cape Girardeau. And just three barges. Very few people on this section of the river. We, of course, never see any paddlers. Just us today. Not even any small fishing boats.

The quotes for today are:

"Look. We're getting passed by a stick." (We were paddling hard, and seemed to be getting nowhere, and watched as a stick floated past. Demoralizing.)

We stopped paddling around 12:30pm for lunch but stayed in our boats. I made my usual peanut butter and jelly bagel, and Jeff had tuna tortillas. He busted open a pouch of dill pickles that were a real treat. But, as we floated along, Jeff said, "Look. We're having a floating picnic." I thought that was funny for some reason.

And my favorite quote of the day: "Fucking wind." It seemed that no matter what direction we were paddling as the river twisted and turned (and it twisted and turned several times today, including a couple of complete oxbows), the wind always seemed to be blowing about 10mph into our faces. It was constant, and definitely slowed us down by a mile to a mile and a half per hour, which is a lot when you think about being on the water for over seven hours.

It only got up to 75 degrees today, and there were a few passing clouds, so not terribly hot. Tomorrow is supposed to be cloudy with les wind. So, maybe our 35 miles will pass more easily.

I sent Leslie flowers today as a thank you for supporting my adventure. I know that it's hard on her sometimes. We're doing some thinking ahead to this coming weekend. Leslie is going to drive to meet us, but there are very few riverside towns coming up. We also heard that a hurricane is hitting Galveston, Texas, today, and the wind and rain will hit us on Wednesday or Thursday (today is Monday?). We need to be paying attention to that, and not camping in some exposed spot if we get hit with several inches of rain.

I talked with Seth a couple of times today. He's obsessing about a 260-acre lakefront lot that is for sale. I'm a little worried that he's not approaching it very rationally, like when he said, "We might put our house on the market in two weeks!" I gave him several suggestions of things to consider. I'm sure I was a bit of a buzzkill, but. . . .

Adventures like this one are so interesting. Moving every day. Working hard every day. Sometimes just feeling like we aren't getting anywhere or making any progress because we still have so far to go. It's hard to stay in the moment sometimes. And really hard to appreciate the moment, and the dozens of experiences we encounter every day.

It's really when we are talking with people who are in awe or our endeavor that it sinks in just a little. This morning, as we were loading our boats, a family stopped and asked about our trip, and then asked to take a selfie with me. They were truly amazed.

And Rocky, our Airbnb host last night, was truly inspired by our trip. By our commitment. Rocky is an Iraq and Afghanistan Army vet. He said he's had a really hard time negotiating daily life since he's been back. He said he struggles a lot. But our trip put a smile on his face and planted a seed of a dream in his heart (his words, not mine).

Beards and Roses (and Jake)

September 22, 2020 - Day 56
River Mile 19 (Upper Mississippi) to Columbus, Kentucky - 36 miles

Total Mileage: 1,387

Jeff: I don't see any place where a parking lot could be.

Jon: That's because you aren't thinking like a car.

What a day!

Definitely our hardest of the last 10 days.

Thirty-six miles is a long way to go in one day. And not only did we not have any current for most of the day, but there were two or three sections where it felt like the water was pulling us backwards for several miles. It was like paddling in pudding.

The first such spot was around mile 19 for the day, where we met the confluence with the Ohio River. For the past few days, we'd been dreaming of the river picking up some speed at the point where the Mighty Ohio joins the Mighty Mississippi. I was sure we'd have a decent current, at least for several miles after the confluence, and Jeff was especially excited to get to this landmark. It was a major milestone for him, one that he'd imagined several times over the past few years of planning and map studying. I was just excited for the water to move faster.

What a fucking disappointment!

Right after the confluence, the river came to a dead stop. We paddled and paddled and just crept along. We saw some tugboats tied off just ahead on the Kentucky side (Welcome to Kentucky), and an hour later we still hadn't passed them. It was exhausting and depressing.

I blamed the gigantic white cross erected on a hilltop on the Kentucky side that starred down at us for over an hour as we crept past. But this no-flow, or sometimes backwards flow, continued for over two hours, and we only traveled maybe four miles.

We'd decided to take a "get out of our boat" lunch break at a boat ramp we spotted on the map around mile 24 for the day, and we finally got there at 2pm. Exhausted and starving, we were both close to just hanging it up for the day. There was a semblance of a parking lot, and a nice grassy area where we could have set up our tents. But . . . it was only 2pm. And just as we pulled up and got out of our boats, it started to drizzle. This is the point when the parking lot quote was coined.

We sat under a tree in an attempt to get out of the rain. I was just happy to be out of my canoe and stretching my legs and sore back. Same old boring lunch.

We started paddling again around 2:30pm with only 10 miles to get to the boat ramp in Columbus. I'd seen a small market on the Google map, so that was our motivation to keep going. To get some food, cold pop, and maybe cold beer. There was also a state park in Columbus, so I had conjured dreams of a picnic table and hot showers. Well . . . it didn't quite turn out that way.

The next five miles zipped right along, at probably five or six miles per hour. But, the last five miles, the river flow came to a dead stop. And it continued to rain from 2pm until we stopped at 5:30pm. Just a constant drizzle. Again, it felt like the river was actually flowing backwards.

Exhausting.

Just creeping along.

We also got into a lot of barge traffic for the last two or three miles, with tugs moving from one side of the river to the other, both in front of and behind us. By the time we pulled up to the Columbus boat ramp, we'd been paddling for nine and a half hours. By far our longest paddle yet, and it was still drizzling.

We were both spent.

We hauled our wet bags and boats up the ramp and I found a dry spot under some trees to set up camp. I wanted to get to the little store I'd seen on the Google map before it closed at 7pm. Some local kids were fishing on the riverbank and gave me directions since we didn't have cell service where we stopped. The store was about a mile away in the bustling center of Columbus, Kentucky, that consisted of the gas station/store, a post office, and a church.

Jeff felt like he needed to stay back with our stuff, so I threw up my tent, stowed my bags and walked into town.

The market, called Beards and Roses, ended up also being a little restaurant. I wish Jeff would've come up with me. I was famished, so I bought a hamburger, a chicken parmesan hoagie, fries, a cold roast beef sandwich, several pops, chocolate milk, and Gatorade. I also bought some groceries off of the few shelves.

The store is owned by Rob (who had a big beard) and Rose. Their nephew Jacob works there too. All super nice, and super interested in our trip. Jacob especially couldn't believe we were paddling the entire Mississippi River.

I was worried that Jeff had started dinner back at our campsite, so I asked if Rob could give me a ride back down to the river. Rob tossed the truck keys to Jacob. I got in the front passenger seat, and Jacob, who had a hard time starting the car, turned to me and said, "Do you want to know how many times I've driven a car?"

Right as I answered, "No, don't tell me," Jacob said, "Only once."

Shit.

That was just as Rob piped up from the back seat, "Jake. Did you bring your permit?"

So, from the front right seat for that long, long mile back to the river, I'm helping Jake drive. *Okay, Jake, turn your blinker on. No, that's your windshield wiper. Okay, easy on the brakes now. Slow down. Okay, now don't drive in the grass, Jake, just keep it on the road. Yup, stay on the right side, Jake. Always drive on the right side of the road.*

Oh, brother. What a day.

I got back to Jeff after being gone for over an hour, and Jeff, who'd said he was going to gather firewood and get a fire started, had just set up his tent and gotten his stuff organized. What he'd been doing for the past hour, I have no idea.

But we found some wood, got a fire started, ate the food I'd brought. And all was well. It's not raining. So, hopefully tomorrow it won't rain either. We decided that we are only going to paddle 25 miles tomorrow. We both need a bit of a break.

Old Guys in Pickups Drinking Beer

September 23, 2020 - Day 57
Columbus, Kentucky, to Mile 912 (Lower Mississippi) - 25 miles
Total Mileage: 1,412

We are camped at a boat ramp about 11 miles from Hicksville, Kentucky. Yes, Hicksville! It's kind of a rocky spot, so we couldn't get any tent stakes to go into the ground. But it's flat and will be relatively easy to get our boats back in the water tomorrow morning.

We opted for a shorter day of 25 miles today, partly because we had such a long hard day yesterday, and partly because we wanted to walk up to Rose and Rob's this morning and have a real breakfast.

Jeff and I were both up and out of our tents by 6:30am. I was all packed up, with my boat loaded on my little trailer by a little after 7am, so I walked the mile back uphill to the store and plugged in all of my devices to do some charging. I also did a little computer work and visited with Rose and Rob until Jeff showed up. We each ordered an amazing bacon, fried eggs, toast, and hash browns breakfast, along with bottomless mugs of the most watered-down coffee I've ever had. I added two instant Starbucks packets to mine, and it still tasted watery.

Rob and Rose were super nice. While we ate, they introduced us to everyone who came in the store. We met Jacob's mom and I made a point of telling her what a great boy he was. We also met Mayor Mary, who happens to be the owner and publisher of the local Hickman newspaper. She asked if we'd take a photo of ourselves when we get to the gulf with a copy of her newspaper that she sent along with us. I hope I remember to do it.

We met a few other locals as we ate.

Rob and Rose almost begged us to let them know when we get to the end of our trip.

Today we had great paddling conditions . . . cool, very little breeze, and overcast. And a little bit of a current heading our way. So, we covered the 25 miles in about six hours, including taking a floating lunch, and a 15-minute break to stretch my legs. Even though 25 miles is low for us on this leg of the trip, it's still pretty much an all-day paddle, and still consists of hour after hour of paddling, and more paddling. There just aren't any shortcuts or ways to go faster. You have to earn every mile. And going further each day just means paddling for more hours. You can't just float your way down the river and expect to get anywhere. Unless you have unlimited time, I guess.

I've been hoping we could finish the Mississippi in 35 days when we put back in 11 days ago, but it's looking more like 42 to 45 days. We'll see. The next two days we'll shoot for 30 to 31 miles per day so that we can get to the town of Tiptonville by Friday afternoon to meet up with Leslie that evening. I can't wait!

I'm woefully behind in my office/computer work. But with inconsistent internet and moving every day and it being pitch dark by 7:30pm and starting to fall asleep at 8:30pm, I'm just not getting much done.

We paddled up to the boat ramp here at River Mile 912, and there were three pickup trucks parked at the top of the ramp, with one old guy sitting in the cab of each truck. They had their windows rolled down so that they could talk between the trucks. They each had a plastic Igloo cooler in the back of their truck, full of ice and Bud Light. Two of the guys asked about our trip as we unloaded our gear and looked for a place to set up our tents. They offered us each an ice-cold beer, which was really nice. And when they all left an hour or so later, one old guy pulled up to our tents and offered us two more beers.

They were just sitting in their trucks, smoking cigarettes, and drinking beers, and tossing their empties in the back. The truck I grabbed my beer out of had eight or 10 empty cans rolling around in the back.

This is a long river. It's a long commitment of time and energy.

It's the grind that is hard.

Paddling for a long day or two is no big deal. But the daily grind, the day after day after day. That's the hardest part. And not letting my mind get too far ahead of itself. And with five weeks still to go, it feels never-ending.

Enjoying the moments. Appreciating the experience. Knowing

that the trip will result in a million and one memories. Meeting really friendly people who all seem to want to participate in our trip in some small way.

This morning Rose gave us two freshly baked sweet rolls for free. A local guy named Josh picked us up on the road after breakfast and drove us back to our boats. The old timers offering us beer. Mary the newspaper editor giving us a free copy of this week's issue. Everyone wanting to help us in some small way.

And all of these little "helps" make all the difference in the world for us!

Free Food for the Homeless Paddlers

September 24, 2020 - Day 58
Mile 912 (Lower Mississippi) to New Madrid - 22 miles
Total Mileage: 1,434

Hmmmm. Where to begin? What to write about?

In some ways, everything and every day runs together. Just one continuous stream. But when I take a moment to reflect, there were so many different parts of today . . . and every day on this river.

Last night, in the tent trying and failing to get a strong enough phone signal to tie in my computer internet, I started feeling anxious about how much work I have piling up. Scores of unanswered emails, and shit that just needs to be attended to.

Then, last night I had a recurring dream that I was back in high school, as a 59-year-old, with finals in a couple of days, and I'm totally unprepared. I can't remember my locker combination, can't find my textbooks, and completely unable to find the solutions I need to solve these simple problems. And I kept telling myself in the dream, "I have a master's degree. I don't need these last few high school credits."

So, my dream and my work anxiety are related, I'm sure.

Jeff was up at 5am, starting his packing process in the pitch dark. It began raining last night around 8pm. We had a great fire going and had to leave it blazing as it started to rain. The rain stopped by 2am, thankfully, but then started again around 5:30am. It drizzled the whole while that we packed up and launched and continued to rain until around 11:30am. Constant. Like a soaking drizzle or very heavy mist.

So, not great visibility, rain, and some wind.

I kept switching back and forth between my raincoat and my wind breaker (which I soaked), and my short-sleeved button shirt. Luckily it was in the high 50s to low 60s so not too awfully cold. Just a miserable paddling morning.

Jeff agreed to going only 22 miles today, so we could stop in New Madrid early and I could find a warm dry spot to get some computer work done. We paddled up to a nice concrete boat ramp about 1pm and pulled our boats out of the water. We wheeled them over to a riverside park next to some picnic tables and brought all of our valuables with us.

Three young local guys stopped over to ask about our trip, and I made a point of telling them that all our stuff was wet and muddy, so we wouldn't be worried about anyone stealing anything.

We packed up our electronics and maps, our two most valuable possessions, and walked through the small town to stop first at a little deli for a late lunch, and then at the New Madrid Community Care Center, that seemed to be an outreach program for people who had fallen on hard times. Jeff asked if we could hang out and charge our devices and do some work on our computers. So, we sat on two cozy couches and worked for a couple of hours in a warm, dry, and comfortable spot.

At one point, one of the nice community care center workers brought us a huge bag of food . . . cereals, canned goods, noodles, and Clif Bars. Jeff said, "No, thank you," but she insisted that they give food to everyone who comes through their doors. The staff also did some calling around to see if there was someplace indoors where we could spend the night. The only option she could come up with involved getting back in our boats and paddling a ways, and then walking two or three miles. So, we opted for setting up our tents by our canoes right at the boat ramp.

I got a few hours of work done, and we're planning another short paddling day tomorrow. So, I'll have a few more hours to work and will feel much better about being able to relax once Leslie gets here late tomorrow night. So, we'll cover around 67 miles over three days instead of two, slowing us down a bit, but again, I really needed to get caught up on some work so that I can stay positive and focused about our paddling.

After hanging out at the community care center, we walked another mile and had dinner at a Mexican restaurant, El Brasco, and then walked back at 6:30pm to our boats to set up our tents as it started to get dark.

Two older women who were out for an evening walk stopped by to chat. They told us all about an 81-year-old paddler who stopped here just two days ago. He apparently gets a big welcome wherever he stops and is paddling with a couple of other guys who are setting up press coverage for him as he paddles downriver.

It's chilly tonight, but not raining, so our stuff should dry out a bit after being wet for the past few days. Tomorrow we'll get to Tiptonville and Leslie!! It'll be another short paddling day. We'll have to start making up for them beginning on Sunday when

Leslie takes off.

Rednecks in Tiptonville?

September 25, 2020 - Day 59
New Madrid to Tiptonville - 17 miles
Total Mileage: 1,451

We are paddling into the deeper south. Sitting at Boyette's Restaurant near Tiptonville, Tennessee. Boyette's is a well-known local favorite that specializes in frog legs and catfish steaks. It's across the street from Reelfoot Lake, a huge lake that was created by a series of big earthquakes in 1811. The earthquakes caused a huge section of land to sink, and the Mississippi River diverted itself to fill the newly created depression, and then reverted to its old course once a new lake was created.

We only paddled 17 miles today. We got to the Tiptonville boat ramp by 1pm. The concrete ramp was all busted apart, so we had to unload our boats, carry our boats and gear over the broken chunks of concrete and smelly dead fish that were strewn about, and then re-loaded our boats on our trailer wheels.

My map research from a few days ago showed a local fire station located right near the boat ramp. Jeff and I were hoping someone would be there when we arrived so that we could ask to safely chain our boats up for the night. Jeff talked with the Reelfoot Lake Inn owner, and he said he'd pick us up at the boat ramp and haul us and our stuff the three miles to the inn.

Adding to the hauling our stuff over a destroyed concrete boat ramp, trying to find the fire hall I'd thought was near the boat ramp and the hope that we had a ride lined up . . . I had to poop. Bad!

Since we'd camped last night at a little riverside public park in New Madrid that was not open to campers, was right in town, and didn't have a public restroom, I didn't go before we headed out. So, I really had to go by the time we got to the Tiptonville boat ramp.

Even though it was a hassle hauling our gear and boats up the stinky, broken concrete ramp, I spotted a Porta-Potty a few hundred yards away at a little park near the ramp. We hauled our boats up the ramp to a soybean field, and then a couple of hundred yards on a dirt road to the main road. I left my boat and walked a few hundred yards in both directions, looking for the fire hall that I was sure was nearby. No fire station. No nothing. Not a single building to be found.

I must've been looking at a Google map of a different town. There was a parking area up on the levee by the little public park, so we pulled our boats to the lot, all the while I was trying not to poop my pants. Just as I neared the parking lot, a truck with a flatbed trailer pulled up, loaded up the only Porta-Potty onto the back of the trailer, and drove off. Things were getting grimmer by the minute.

An old timer, Glen, pulled up in a beat-up pickup, and we chatted for a bit. Glen offered me a big bag of freshly picked "Tennessee Tomatoes." Like 30 tomatoes!! I couldn't refuse them. It was such a nice offer.

Then another old timer pulled up in his truck. Mike. Both Mike and Glen had a Tennessee accent that was so thick I could barely understand them. We told Mike and Glen that we were paddling the Mississippi and had been planning to lock up our boats somewhere in the park, and they both said, "No way."

Mike offered to drive home, grab his flatbed trailer, and haul us and our stuff to the inn.

Super nice. Couldn't have been a bigger redneck. Mike looked and sounded like an older version of Larry the Cable Guy.

So, when he returned with his trailer, we loaded everything up. I rode with Mike. And Jeff rode with Glen to a Dairy Queen to buy lunch while Mike and I unloaded our gear and boats. Mike offered anything he could do to help us these next few days. Just call. He said he'd always dreamed of hiking the Appalachian Trail, but he's too afraid that his "bad heart" wouldn't make it.

We did laundry at the inn, and I got a few hours of work in.

The paddling today was pretty uneventful. Overcast. Maybe low 60s. Perfect for paddling. The only excitement came when three huge tows, one 35 barges, and another of 24 barges, and a third, all passed us within 30 minutes of each other. The water got super wavey, and I ended up too far from shore, and really got knocked around.

Two- to two-and-a-half-foot waves came from all directions, and in not predictable rhythms. I took a little water over the side of my canoe, but any one of those waves hitting me wrong would've filled my boat and rolled me over. I was pretty tense and focused for about 20 minutes or so. Other than that, there were some periods of slow, stagnant water, and some periods of faster moving water.

For dinner tonight, we walked about a mile and a half to Boyette's Restaurant. Both Mike and Glenn said it was the best place around and that we had to go. So, we did. No outdoor seating, but the wait staff wore masks, sort of, and most of the tables were empty when we arrived. We're both really wary of eating indoors with COVID-19 still on a rampage, especially in the south.

I had catfish steak and hush puppies and Jeff got the "seafood platter," which consisted of catfish, shrimp, and several huge frog legs. (We both remarked that that was an odd definition of "seafood.") By the time we were eating, the restaurant was totally filling up with non-mask wearing Tennessee locals. Groups and groups came in with no masks on, and we started noticing that most of the wait staff now had their masks pulled partially down or hung them around their chin.

Jeff said that he was starting to freak out. We'd let our guard down and were now in the middle of an indoor super-spreader event. Neither of us finished our meal before we got our checks and walked out. COVID-denying rural Tennessee. In retrospect, we never should have gone in to begin with.

I have to say that the catfish steak was delicious, and my bite of one of Jeff's massive frog legs was moist and tasty.

It's 7:45pm and we're sitting out in front of our motel rooms, getting chewed by mosquitoes, and waiting for Leslie to arrive.

Another nice thing that happened when we were loading up this morning: Jerry, a Vietnam Vet, stopped by and told us all kinds of local history of the town of New Madrid—the earthquake of 1811 and 1812, and river stories about various paddlers that he's met over the years. He offered to get us breakfast, or run us into town, but we were already packed up and almost ready to launch.

Jerry told us that when the 81-year-old paddler was here a few days ago, Jerry loaned him his truck to run errands and buy groceries and beer. Another old, nice, retired guy hanging out by the river, just waiting to lend a hand.

Leslie to the Rescue

September 26, 2020 - Day 60
Tiptonville to Caruthersville - 26 miles
Total Mileage: 1,477

Leslie is here! That's the highlight for sure.

She dropped us off at the boat launch near Tiptonville and we were on the water by 8:30am. And then she met us in Caruthersville where we took out at 3pm. I wish she could do this every day!!

We were all up early this morning. No packing the boats because we just carried stuff for the day. We drove into Tiptonville around 7am to the Bean Me Up coffee shop. It's a cute little coffee shop that just opened a week ago.

Mike met us at the motel at 7:45am with his trailer. So awesome of him. Leslie said that Mike makes her want to be a better person. He helps, unconditionally. Mike has done motorcycle trips all over the country, and he says he's been in plenty of situations where he's needed help. So, he just wants to help out other people that he meets.

When we were unloading our boats and getting ready to launch, Mike was talking about fishing for catfish, and he said, "We call them Looky-up Fish," because they sit on the river bottom with eyes on top of their heads.

It was overcast all day, and sort of cool, which was fantastic. A far cry from the 82 degrees and sunny that was predicted. We got into a lot of wind around mile 10 and paddled our hearts out, getting nowhere. I hate that.

The wind got stronger as the day wore on, but the "S" turns and the oxbows in the river meant that sometimes we were paddling into a strong headwind, and other times it was at our backs.

Around mile 15, we were passed by four large barges, all within a half-hour of each other. And that was the only boat traffic we saw all day until just before we hit Caruthersville.

Around mile 20, the river turned sharply to the left and the current seemed to suddenly pick up. The wind was blowing hard in my face, and the result was some really fucked up water—two- to three-foot waves coming from all directions, with a swift current and lots of whirlpools popping up all around. I was nervous and hyper-focused on staying balanced and trying to take the waves head-on. I hate being in this little 13-foot canoe in this kind of water.

I was actually talking with my mom on the phone through my ear-buds when I got into the scary water. So, for 20 to 30 minutes, I tried to talk as calmly as possibly while I talked loudly over the wind and scanned for whirlpools and cross waves. I didn't dare try to turn around to see where Jeff was. I just kept paddling hard.

We stored our boats behind the Grizzly Jig fishing supply store. The owner's son showed us a place where we could lock them to a big truck. I'd called them earlier in the day to see if they'd be willing to help us out with storing our boats for the night. The owner originally suggested we keep them in the store overnight, but they aren't open tomorrow, so we locked them up outside.

On the way back to the inn, we stopped at The Pit, a barbecue joint in Tiptonville, and had amazing smoked chicken, pork, and a full rack of ribs with baked beans, amazing potato salad, and an order of deep-fried okra. We brought some beer with us and had a nice picnic in their screened-in front porch.

We spent the evening looking at maps, journaling, making plans for tomorrow, and talking with Leslie about home stuff, since she's leaving to head home in the morning. It's been so nice to have her here. We need to get back into our 30-plus-mile days if we are going to finish in four weeks.

One day at a time.

Beach Fires

September 27, 2020 - Day 61

Caruthersville to Mile 820 on the Lower Mississippi Map - 25 miles

Total Mileage: 1,502

Sitting by a nice beach fire. It's finally starting to cool off. When we got here around 4:30pm it was hot and muggy. My shirt was damp with sweat. But it's calm. The wind we battled all day long has totally calmed down. A peaceful evening.

We are both beat. It's only 7pm, but this is the great part of each day. Paddling is done. Tents are set up. Sleeping bags and pads rolled out. Gear is all packed away. And I don't need to think about tomorrow until tomorrow.

We were in the car all packed up and heading back to Caruthersville by 7am. I got to sit in the car with Leslie for a while as Jeff packed his boat down by the water. I wish she could stay longer. I tried to make having her come to visit again in another two weeks sound really fun. But it would be a minimum of nine and a half hours of driving each way. So, I think it's a big "no way."

We pulled out at 8:45am. Lots of barge traffic today. This seems to be more the norm on this part of the river. The first five miles we averaged 5mph, but then the wind picked up and blew a steady 10 to 15mph directly into our faces for the rest of the day. The temp got up into the low 80s and the skies were mostly clear and sunny. But the wind really slowed us down. For a few hours, we averaged only 2mph and that was paddling hard and non-stop. Stopping paddling meant drifting backwards. The wind was so strong that it was piling up waves and swells and pushing them backwards upstream.

We stopped around 1pm and Jeff actually got out of his boat, a rare occurrence these days. We had our lunch on a sandbar. We had the rest of our barbecue from last night. I laid in the warm sand for a bit and closed my eyes. I could've fallen asleep very easily.

It was hard to get back in the boats to keep going. Just a hard slog. Lots of boat traffic as I mentioned. A few of them threw up some big crazy waves. One small tug with only two barges headed out of the shipping channel and right for us. We had to paddle way over on the side of the river to get out of his path. We were like, WTF? But the tug was pushing these two barges to hook them up with some other already parked and tied off barges along the side of the river.

At one point, there were three-foot waves coming at my boat, but they were pretty consistently spaced so that I could just paddle straight up and over each one. On flat water, when my boat is full with me inside, there is only about eight inches of "free board," the distance from the water line to top of my boat. So, paddling in three-foot waves makes me nervous.

We are both longing for a day or two with the wind at our backs. We just haven't had a wind out of the north for a long time.

I called Seth and we talked for a while as I paddled. He heads back to work tomorrow. They are both worried about how Arlo will do with Brittany, their sitter.

We'd hoped to go 29 miles, but only got in 25. Getting in only 30 miles per day, consistently, is starting to feel out of reach with this wind. We gave up on the idea of 40-mile days a week ago. The river level is unusually low this year, so there is often little to no current, and the constant wind out of the south just grinds us to a halt if we aren't constantly paddling.

I'm just trying to focus on the moment, on the hour, on the day. But I can't help thinking ahead. Mileage-wise, we are about a third done with this final leg of our trip. Four-hundred and 12 miles down with about 820 miles to go. Today was day 15 of this leg, so that means about 30 to go. Or, if we could start averaging 30 miles per day, that would be 27 and a half more days. I do these calculations in my head all day long.

Just gotta keep plugging along.

Right now, it is peaceful. Hoping to get the Sunday night Packer game on the radio in a few minutes. I need to enjoy the moment, because come tomorrow, we'll be paddling all day again.

Rain, Rain, Go Away

September 28, 2020 - Day 62
Mile 820 to Mile 796 on the Lower Mississippi Map - 26 miles
Total Mileage: 1,528

8am.

It is blustery this morning!

Wind blowing 20 to 25mph with gusts to 35mph.

Heavy rain with a little lightning thrown in.

So, we're just laying in our tents. Riding it out.

Our goal is to be on the water by 10am and just paddle until 5 or 6pm.

The water looks fast this morning.

That makes me nervous.

Our first really bad weather in the last 16 days. We'll pack up wet and hopefully dry out tonight.

6:52pm.

Well, we're drying out.

Tents are set up and drying. Rainwear is drying. It's still breezy but it seems to be dying down.

It's getting chilly.

The sun just set, so I need to finish journaling before it gets too dark.

We're camped at a boat ramp at Mile 796 of the Lower Mississippi map, right next to a massive cotton field. There is one house on stilts nearby. We haven't seen any houses along the river in the past two weeks.

We were on the water by 9am, once the heavy rain and lightning stopped. Jeff moved extra fast packing up, due to the rain. I fired up my MSR Reactor stove and heated water for coffee and oatmeal during a little lull in the rain. But as soon as we started paddling, it started to rain hard, and kept raining until 11am. I had my full rain gear on but was still soaked to the skin.

It was miserable paddling in the rain, but we were moving, and had a nice wind at our backs.

Finally!!

We were definitely going 5+mph for the first hour.

About an hour in, we passed two guys in a canoe! The first people we've seen on the river since we got back on the river 16 days ago. Two guys in a small two-person canoe. It didn't look like they had much stuff, though the sides of their canoe were only a few inches above the waterline. They were using old school straight-bladed wooden paddles. We paddled past them where they had pulled off the river on a sand bar to pee. They paddled in perfect synchronization. Hard and fast. Almost twice as fast as our paddling pace.

They said they were on Day 39 since leaving Itasca State Park. Today is Day 62 for us! The weird part was that as we passed them it was pouring rain, and one guy had a heavy flannel shirt and shorts on, and the other guy had on a light sweatshirt and shorts. Something didn't seem quite right. There was no visible gear in their canoe, so whatever little gear and food they had laid down below the gunnels.

It was a good paddling day, but with the twists and turns in the river, the wind helped us sometimes, and hurt us sometimes. So, we ended up averaging just under 4mph for the day. We'd hoped to go 30 miles, but we were both ready to stop at 4:30pm, at 26 miles, especially since we had a boat ramp to pull out on. Boat ramps make unloading and loading so much easier.

There was a guy in an SUV parked on the ramp with a bright yellow kayak strapped on top. Dave was waiting for his wife. She apparently started in St. Cloud, Minnesota, and he meets her every night with the truck, picks her up, and they either stay in a motel or camp together. So, she doesn't paddle with any gear.

As we were setting up our wet tents alongside the boat ramp's gravel parking lot, she paddled up. After they'd strapped her boat to the top of Dave's car, she came over and said, "You guys willing to have a paddle buddy tomorrow?" I wanted to die.

Hell no.

Jeff, of course, said, "Sure."

So, I guess we have a complete stranger for a paddle buddy tomorrow. Fuck.

I was annoyed with Jeff for so eagerly agreeing. I said to Jeff that when it's time to leave in the morning, if she's not here, I'm not waiting. She also said something about wanting to go 40 miles tomorrow. So, if that's the case, she can just paddle on ahead. Because there is no way we are going 40 miles tomorrow, or probably ever. Twenty-six miles was hard enough today.

So, we'll see how having a "paddling buddy" goes. The last time we had one, his name was Bob and we all know how that turned out.

While we were setting up, an older couple pulled up and asked us a bunch of questions and offered to give one of us a ride to a little market about ten miles away to buy something cold to drink. So, I jumped in back and breathed in cigarette smoke the entire 20 miles round trip. The store only had junk food, but I bought some beer, pop, and some pre-packaged cinnamon rolls for breakfast tomorrow. Another case of super nice locals willing to help complete strangers.

By the time I got back, Jeff was gathering firewood. So, we had some of Seth's smoked trout, the rest of our brie (from Leslie), crackers, and a big pickle I bought at the store. Great dinner. And two giant cans of Budweiser.

It was a good day and it's going to be chilly tonight.

Paddle Buddy, Not

September 29, 2020 - Day 63
Mile 796 on the Lower Mississippi Map to Randolph Boat Ramp - 30 miles
Total Mileage: 1,558

Oh, Rebecca. We hardly knew you.

What happened to you, Rebecca?

Our paddle buddy is no longer with us.

It was chilly last night and early this morning. I went to bed in my skimpy little sleeping bag with dry socks, a lightweight thermal shirt, and my boxers. I was in the fetal position for most of the night, trying to stay warm.

Heavy dew equals packing up a soaking wet tent.

We had the sweet rolls I bought yesterday and coffee while we packed up. Jeff's alarm went off at 6am, but he didn't get out of his tent until 6:30am. Late start for Jeffie. He vowed to be on the water by 8am, but something must've tripped him up. At 7:50am, I looked over at his cluster fuck of shit scattered around and asked, "Are we still on track for an 8am launch, Jeff?" He confidently replied, "Yes, we are!"

Then our paddle buddy arrived, and Jeff slipped into an even slower gear.

Rebecca was plenty nice. But, Lordy, could she talk. She talked and talked, even as my boat drifted way. She just kept on talking. Jeff and I took turns paddling close enough to engage in conversation with her. But it really didn't matter if we were close by or not. She just kept prattling on.

Rebecca was a bit of a know-it-all. She would say things like, "I've been told to always paddle down the middle of the river." She told me that three or four times, and every time I'd reply that the middle of the channel is where the big barges go, so I stay out of the middle. She told me, "You need to work on that." Rebecca had all sorts of advice. She was the type of person that would ask a question and then not wait to listen to the answer.

And she LOVED the idea of paddling in a group. She wanted to be an instant "team member."

She told us that she'd paddled 72 miles in a single day three days ago. That was the day that Jeff and I busted our asses to go 26 miles. No way.

Rebecca said she averages 40 miles per day but was hoping to get it up to 50 miles per day.

She said that someone told her that it only takes nine or 10 days to paddle from St. Louis to New Orleans.

I said, "That would be 90 to 100 miles per day."

She laughed and said, "Oh, people do it."

I said, "I'd have to literally paddle 24 hours to go that far in a day."

So, Rebecca was full of all kinds of nonsense that made me wonder about everything else she was saying.

Anyways, like I wrote, Rebecca was nice and really liked paddling with other people. She paddles with an empty kayak, because her husband meets her at the end of each day. And she doesn't need to stop earlier in the day to set up camp and cook, since they have a car and can drive to hotels and restaurants.

The first 25 miles zipped by today. I've never written that before. We had the wind at our backs, and a bit of a current.

But then, the wind picked up and whipped around and started blowing 15mph into our faces, and it was right at a tight turn where the water was already doing crazy, choppy things.

I almost got flipped over by a whirlpool of swirling current. And for the next 90 minutes or so, we plowed right into the wind and chop. Intimidating in a small canoe on a big river.

Mile 29 was where Jeff and I had originally thought we'd stop for the day. But since we'd been moving along so well, we both decided to go further. We were approaching a massive wing dam of piled up boulders that stretched about 2,000 feet into the middle of the river, which made the rest of the river move faster around a big bend. Rebecca started paddling her kayak into the waves to go around the wing dam just as a 35-barge tow was coming up and around the wing dam in the opposite direction.

I was already nervous about the big waves and knew that the waves from the barges would make it way worse, so I made a beeline for the boat ramp at mile 29. These big waves coming from all directions scare me.

Jeff followed. I was ready to suggest that we just take a break and regroup before going any further, because I knew that Jeff was excited about seeing his wife Chris tomorrow. But he paddled up next to me and said, "Fuck it. I'm done. Let's get off the water." He didn't have to ask me twice.

Rebecca paddled out and around the wing dam and we haven't seen her since. Her husband was planning to pick her up somewhere further downriver.

So, we were setting up camp by 2pm!! The sun was out. We put our tents behind some shrubs out of the wind. Jeff struck up a conversation with a very old couple who were parked at the ramp and watching the barges go by. Billy eventually offered to take us into the town of Munford to get some food. As we got to his truck, he said, "Do you guys have masks? You gotta wear your COVID masks."

Right on, Billy.

So, we organized our gear a little and climbed in the back of their pickup. All four of us wearing masks. It was around 15 miles to the

Burger King in Munford, Tennessee. Billy used to be a tugboat pilot, and he told us several stories.

Super cute couple. I really enjoyed them. They said they help out all kinds of people, and that they really believe in paying it forward. So cute. Really made our day. And the Burger King food was fabulous.

The boat ramp where we are camped is super busy. Several loud pickup trucks pulling in and out. A pickup just roared past with a confederate flag waving in the back. We are in Southern Tennessee.

Scariest Day Yet

September 30, 2020 - Day 64
Randolph Boat Ramp to Shelby Forest Boat Ramp - 16 miles
Total Mileage: 1,574

What a day!

Definitely the hardest and scariest yet!

It makes me wonder about being on this massive river in a small solo canoe.

Jeff was up before 5am, starting to pack. I could hear him puttering around in the dark.

We'd planned to paddle 31 miles to a marina right in downtown Memphis. Our goal was to be on the water by 7am so that we could get to Memphis as early as possible.

I got up around 6:15am. It was still pitch dark. The sun rose around 6:45am, and we were on the water by 7:20am. The wind was coming out of the south and directly in our faces, and by 8am it was blowing a consistent 20mph non-stop for the rest of the day.

We'd decided that if the wind was bad, we'd stop at a boat ramp at 16 miles. And it was worse than bad. The wind blowing in the opposite direction of the current caused lots of waves. And, as we'd found out on previous days with a south wind, around 90-degree turns the waves just pile up on top of each other. Add in three- to six-foot-tall barge waves and we were fucked.

We ended up paddling for eight and a half hours and only went 16 miles, and that distance was an epic struggle. It was 16 miles to a boat ramp, and no place for Chris to meet us before that point, so we had to go that far into the wind.

Besides paddling as hard as I could and literally creeping along at a mile-per-hour pace for most of the day, there are four parts of the day that warrant a brief description (presented in chronological order):

(1) About five miles into today, the wind started to pick up, so we got behind a rock wing dam to take a break and let a big barge heading upriver pass us. As the first one was passing, I noticed a second huge tow of barges not bar behind, so we decided to wait for that to pass as well. The wind was piling up waves as it pushed against the current, the water rushing around the wing dam was really moving fast and the big tow was throwing huge waves. I waited about five minutes after the second barge passed and pulled out from behind the rock wing dam and into the current to check out the waves and river speed.

Jeff said, "This looks pretty bad, Jonny."

I was nervous.

If Chris hadn't already been on her way to meet us, I think we both would've said, "Fuck it," and just re-set up camp at 9am, to wait for the wind to die down.

So, I paddled out into the current, and the waves were way bigger than they'd looked from a distance. At least five feet tall and coming from every direction. By then I was committed and scared as I got swept around the outside of the wing dam. I rode up and over several big and cresting waves and tried to keep my bow heading directly into them. Jeff was right behind me.

I took the top of a wave over the side of my canoe.

Jeff yelled that he was going to head towards the middle of the river, to the right, to get away from the mess that the wing dam was causing. He thought it didn't look as bad in the middle of the river. I wasn't interested in trying to get further away from shore and didn't have enough control over my boat to get there anyway. I knew that I just needed to keep my bow into whatever big waves were coming towards me.

I was sure I was going to flip.

After a few minutes of doing all I could to keep my boat from turning sideways into five-foot waves, I saw a chance to peel off to the left and paddled like crazy. It was another few minutes before it settled to three-foot waves. But I was out of the worst, and we were on separate sides of the river. I've never paddled in waves that big in an open canoe before.

(2) A couple of hours battling the wind later, there was a big sandy beach (island) ahead, maybe a mile away. I decided to head for it to stop and take a break. I'd been paddling for four hours already. As I turned towards the sandy island, I could see a sandstorm being kicked up by the strong winds, and then the water got crazy.

The wind was gusting about 25mph, and large swirling eddies were forming around the upriver side of the island, which were screwing with the downriver current. All of which was causing a whacky chop of waves about three feet tall and coming rapidly from all directions. Again, I was sure that I was going to flip at some point, but just kept paddling as hard as I could to get closer to the sand island once I flipped over.

It seemed never ending, probably because I was only moving one half-mile per hour.

Exhausting and nerve-racking.

But after almost an hour, I finally got to shallow enough water to jump out of my canoe.

And no sign of Jeff.

(3) Where I jumped out, I considered putting my boat on wheels and trying to drag it through the sand, since the island was about two miles long. I figured I'd move faster than continuing to paddle. But my loaded boat was way too heavy to pull through the soft sand. So, I grabbed my bow line and walked along the shore in shallow water, and into the strong wind, and just towed my boat along for a good hour. At least I wasn't scared, and on the verge of tipping over in the river. Even though I couldn't see Jeff on the other side of the quarter mile-wide river, I figured he'd eventually see me and paddle back across to join me. The sand was really blowing in my face and nose and eyes. It was slow, hard going, but again, I felt safer than paddling in this mess. I found out later that Jeff was sure I'd flipped my boat back by the wing dam and was worried that I'd drown.

(4) Jeff eventually spotted me and crossed back over. We started paddling together again around hour seven and going as hard as we could. We were maybe going a mile per hour. Finally, after eight hours of this bullshit, I pulled off onto the sand and started unpacking my canoe about a mile before where the boat ramp was supposed to be. I could see on my Google map that a road was only about a half-mile from the river. So, I pulled my boat

up on the muddy sand, and carried a load of packs across a few hundred yards of beach, through the woods, and eventually hit a gravel road. It was not the most rational thing I could have done, but I was so over paddling in this wind, and decided that anything would be better than staying on the water.

Jeff thought I was crazy and continued to paddle on at a snail's pace. After two roundtrips, I eventually got most of my stuff to the road, and then decided it would be too hard to haul my boat through the woods. So, I got back in my almost empty boat and tried paddling again for about a half-mile until I got frustrated again at barely moving at all.

So, I pulled my boat up on the sand again. Strapped on the trailer wheels and pulled my boat about a quarter mile to the road. By the time I walked back down the road to gather up all of my bags and gear, I was completely exhausted.

A very long, hard and scary day, with little to show for it but aching shoulders.

Chris ended up meeting us at the boat ramp around 5pm. She had a hard time finding us, since the spot didn't have a name to plug into Google maps, and cell service was bad. But she found us, and we were at a hotel in Memphis by 6:45pm! What a day!!

Day with Chris and Bass Sporting Goods

October 1, 2020 - Day 65
Shelby Forest Boat Ramp to Memphis - 16 miles
Total Mileage: 1,590

What a contrast to yesterday. We paddled 16 miles in just three hours today, compared to eight and a half hours yesterday. Perfect paddling conditions. Light breeze at our backs. Low 70s. Pretty flat water. Really nice.

We could have paddled 40 miles today, but all of our stuff was back at a hotel in Memphis, so we stopped at the City Marina in Memphis, and were done paddling by a little bit after noon. So nice. I wish every day was like today. Good current. Barge traffic wasn't too bad.

We stored our boats on a dock at the marina and walked to the fabulous glass pyramid of Bass Pro Shops. An amazing building. I have been looking forward to getting to Memphis to buy a warmer sleeping bag, a new cotton t-shirt for sleeping, a warm poly-pro hoodie, a fuel cannister, and a dry bag for my new sleeping bag. We had lunch at the restaurant at the store, and then took an Uber to our hotel.

I picked up some food at a Walmart, and a new book to read. I just can't continue with Mark Twain's *Life on the Mississippi*. Super boring. Sorry, Mark. I've been trying for 18 days, and just don't want to read it anymore.

So, tonight it's journaling, office work, repack everything (tent, sleeping pad, etc., are draped all over my room, drying), and then Jeff bought steaks to grill for dinner at the hotel's outdoor grill. Brilliant idea. I might watch some Thursday night football tonight too. It was a much needed, relaxing day. Tomorrow we'll set our sights on 30 plus miles.

Facing Fears and John for Obama

October 2, 2020 - Day 66
Memphis to Norfolk Star Boat Ramp - 26 miles
Total Mileage: 1,616

Great day! Started off a bit lame, but we got in 26 miles. Great paddling conditions. Highs 60s; light wind from the north; decent current. If we hadn't gotten such a late start, we could have easily gone over 30 miles today. All in all, a good paddling day. I'll get the negative thing off my chest first, and then write about four progressively more positive things.

Negative: Jeff had two days at a hotel to go through his stuff. I know he did it. But it wasn't until we were at the marina at 8am that he realized the closure on his map case was broken. I suggested he use a dry bag, or a zip-lock bag, or duct tape, but he was convinced that he had to have a new one from Bass Pro Shops. So, it took over an hour for him to call the store, get an Uber ride, purchase, and return. And . . . they didn't have what he wanted after all!! For some reason, I really let this delay drive me crazy.

Okay, I got that out of my system. Now, for my positive things, in increasing awesome order.

Positive Thing #4: I get to use my brand new 20-degree sleeping bag tonight, and I am so excited. It looks very cozy laid out in my tent waiting for me.

Positive Thing #3: Facing fears. This is sort of a throwback to the journal entry that I made a couple of days ago. I've been a bit freaked out about the wind and waves and big water, to the point of wondering whether I want to go any further. The water level of the Mississippi River is significantly lower than usual (past years at this time). This makes the river narrower, so we have to paddle closer to the huge barge tows as they pass us. This also make the water shallower, which causes the waves to be larger. Three people in the last two days have told us about several people who died on the water near Memphis, by being pulled underwater by fast currents and undertows.

Guys who were doing dredging on the river, and staying at our hotel in Memphis, said we were crazy to be out on the river now in our tiny boats, especially with so much barge traffic. But I've just decided to be as smart as possible: give the wing dams a wide berth, especially when the wind is blowing from the south; stay closer to shore; don't paddle in big winds. Just be smarter. And take it one day at a time. So, I felt really proud of myself for getting back on the river with some renewed self-confidence.

Positive Thing #2: Our Uber driver, John, took us from the hotel to the marina this morning, where we'd tied our boats off on a dock by the marina office. We talked about a lot of things that quickly turned political. I talked about how much I love the Obamas, and how I miss them and their decency and intelligence and humor in this screwed up era of Trump. John referred to him as Barrack and said that he loved him too. He also said that he'd love to see Michelle Obama naked.

John is a black man from Mississippi. He is in his mid-60s I'd say. He said that he grew up in Mississippi in the 1960s and witnessed the beginnings of the Civil Rights movement. At one point, John said, "I lived through the Civil Rights movement in Mississippi as a black man, and I can't tell you how wonderful it is—and I'm just being honest—to hear two older white men talk about how much they love Barrack. It makes me so happy. Thank you so much." That's what John said.

Positive Thing #1: I woke up this morning to the news that Trump and Melania have contracted COVID! (Sick thought, I know, but this is my journal and I'm being honest.)

The other notable thing today, was that five northbound barges passed us within a single hour! Lots of barge traffic today. But no real big waves, so that was good.

I think we only have about 20 more days. I'm ready to begin the countdown.

Wayne
October 3, 2020 - Day 67
Norfolk Star Boat Ramp to Walnut Bend Boat Ramp - 33 miles
Total Mileage: 1,649

Well, the fact that the above notations say that today we paddled 33 miles from one boat ramp to another boat ramp pretty much sums up the day. Today we were just putting in the miles.

Trees.

We paddled past lots of trees.

Barges.

Lots of barges passed us, heading both up and down the river.

Pelicans.

After seeing all of those pelicans a while back, we hadn't seen any until today. A few loners, and a few heading south. I also saw two eagles chasing each other around, probably flirting.

This morning, everything was soaked from a heavy overnight dew. But it was so nice to eat my oatmeal and drink my coffee while sitting at a picnic table. Jeff didn't start puttering around until a little before 7am. We set 8:30am for our departure time, so I did some computer work while Jeff jacked around with his stuff. I think we ended up pushing off around 9am.

The first 15 miles zipped right past. We had some current, and no wind to speak of. We passed a casino and parking lot at mile 11. I'd been planning to stop and check it out but ended up paddling right on past. I ended up regretting it because that turned out to be the only point of interest along the way today.

We were going to stop at a boat ramp around mile 22 for lunch, hoping for a nice one, like where we camped last night, but we paddled right past the spot noted on the map, and never saw a place to pull off. So, we crossed to the other side of the river, and had lunch on a sand bar. I just like getting out of the boat once in a while. After sitting for four hours, I need a stretch break.

I talked with my brother Aaron on the phone. He was home making his traditional Saturday morning breakfast.

I forgot to mention that last night at the river park where we camped, an old guy walked over from the parking lot to where we'd set up our tents to visit. Wayne was super nice. He asked if he could get us anything, and Jeff said, "beer," before Wayne could finish his question. So, about a half-hour later, Wayne returned with a six pack of Bud Light.

Sitting at a picnic table in a park on the banks of the Mississippi River, in the state of Mississippi, drinking Bud Light.

Our goal was a boat ramp at 33 miles today. That's more than we've paddled in the past ten days. The boat ramp was muddy and busted up. Tough to get our gear out of the boats, and the boats out of the water, with silty mud up to our calves. At the top of the ramp was a gravel parking area, and Porta-Potty, and nothing else. We picked a spot off to the side that had a little grass under some trees. Sort of lame after a long day of paddling. But once the tents are up, and dinner is cooked, a little fire burning, and some warm whiskey in our plastic tumblers, it's all good. I know we still have a lot of days to go (20?), but today is done. We got in some good miles, and tomorrow there is supposed to be a slight wind out of the north. Nice!

Later in the evening, two guys pulled up to put their boat in the water to do some night fishing. I asked them if there were any riverside towns coming up, so I could have something to look forward to, like a cheap motel or an Airbnb. In 15 miles is West Helena, but its only 15 miles, so I suggested to Jeff that we stop there to walk around a bit tomorrow, just to break up the paddling a bit. Then around mile 25 the Army Corps map shows Friar Point. That might not be anything, but I hope it is something. Some reason to stop there for the night. We don't have any internet tonight, so I can't look ahead.

We want to average 25 to 30 miles per day from here on out. At that pace, if we decide to take the Atchafalaya River to the gulf, we have about 20 days to go. God, when it gets down to 14 more days, and 10 more days, we'll be able to smell the end. Speaking of "smell," the inside of my tent smelled like pee when I set it up this afternoon. Uh oh.

Oh, wait. One more thing. Around mid-afternoon today, a guy in a loaded Lund skiff, maybe 14 feet long, pulled up next to us. Ron Miller said he'd paddled from Lake Itasca to St. Louis, and then decided, "I'm too old for this shit." So, he bought the Lund, and is heading to the gulf in that. Today he was going from Memphis to West Helena, about 70 miles. Must be nice. He looked younger than us. Too old for this shit?

Catfish Racists

October 4, 2020 - Day 68
Walnut Bend Boat Ramp to Friar's Point Boat Ramp - 25 miles
Total Mileage: 1,674

Camping at three concrete boat ramps three nights in a row. That's a record. Kind of a lame record, but it's something. I need to go back to the father and son catfish fishermen who we met last night. Once they'd set their lines and hooks, and come back to the boat ramp, they saw our campfire, and came over to offer us two cold beers and to shoot the shit.

I've never been in a conversation with so many uses of the word "nigger," "nigger girl," "darky town," and "nigger hooker" in my life. These two, fat, white racists were gross. The dad was the worst. And if I hadn't been worried that they were each packing loaded pistols, which I'm sure they were, I would have asked them to leave.

Two quick examples (you can skip this next section, Mom):

> *Yeah, my brother was at a titty bar and was getting ready to get a blow job from a mother-daughter pair that didn't have one full set of teeth between 'em. And they were dirty. The only clean part of them hookers was the four titties that my brother was sucking on.*
>
> *Yeah, you two go into West Helena tomorrow and get yourself some black pussy. There're hookers standing on every corner. Some real nice nigger girls.*

And it went on like this for an hour.

I've been thinking today that I don't have a vocabulary to describe people like this, and, actually, several people we've met as we've paddled into the Deep South. We have been warned by at least three different white men to not go anywhere near West Helena . . . "It's too dark." "I'd never go there without a gun." "It's dangerous there. Keep an eye on your stuff." All because it's a predominantly African American community in the Deep South.

No vocabulary to describe these men who are offering us assistance, beer, rides to a store to buy food, cold water. Who are interested in our trip. But who are the most racist, right wing, Trump-loving people I've ever encountered. It is impossible for me to say that they are nice people. Or super friendly. Because if Jeff or I were black, we wouldn't have gotten the same reception.

One more example. . . .

We decided that we needed to set more fun goals along the way and stop at the few places that we can along this portion of the Mississippi just to inject a little more fun, and to have things to look forward to.

So, we decided to pull off at West Helena, about 15 miles into our paddle today. There was a slight breeze at our backs for most of the way, so it was pleasant paddling. But, paddling 15 miles still takes time, and we got there around 12:30pm. The channel we had to paddle up into to get to the boat ramp shown on our Army Corps map was being dredged, so there were several working boats and a big dredging barge clogging the narrow channel.

We weren't sure where we could pull off. We tied off on a small floating dock, next to a big boat that was also tied off there. Robert O'Brian, the boat pilot, was another indescribable, super nice, super helpful, super friendly, super racist. One of his remarks was, "I'm Irish. O'Brian is my last name. When my great grandpa came to this country, he worked for 10 cents a day. That's slavery, too, you know. And he didn't complain about it."

Another Robert line: "See that Army Corps dredge barge over there? Everyone working on it is black, and I'm sure they're all fresh outta prison." So many places to go in response to those ignorant remarks, like, "Did your Irish great grandpa come over in chains against his will?" Or, "Did his wife get raped, and his children, who would be your grandpa, get sold off to the highest bidder?"

Fucking ignorant fucker.

We couldn't get away from Robert O'Brian fast enough.

By the way, West Helena, Arkansas, is a small town that sits along the west shore of the Mississippi River and is considered by many to be the birthplace of the blues. They host the world-famous King Biscuit Blues Festival every year in October and have interpretive signs all over town pointing out famous, and now vacant, juke joints and other venues. We were there on a Sunday, so nothing was open, which was a bummer. And, the Southbound Tavern and Restaurant, where we'd hoped to grab some lunch, was also closed. The owner saw me peering in the window and came out to say hello. After hearing our story and telling us that the next nearest restaurant that was open was a three-mile walk away, she offered to make us two Caesar salads to go.

I would love to go back to spend more time in West Helena, Arkansas. The racist father and son catfish fishermen couldn't have been more wrong.

I walked to the Dollar Store and loaded up on junk, including Captain Crunch cereal, chips, cans of Vienna sausages, pop, and nuts. All shit, but it was all they had. And who knows when we'll be at another town.

The wind had picked up during our hour and a half visit to West Helena and was blowing 15 to 20mph from the north. So, the right direction. But the wind was creating big cross-waves and swells that made me nervous. We crossed to the Mississippi side of the river, since that's where our boat ramp stopping point was.

But for the next 10 miles, the wind and waves and swells really kept me alert, cautious, and a bit scared. All exhausting feelings to maintain. I paddled the whole way only 20 to 30 yards from shore, so it was a lot slower, but I felt safer. Jeff, in his sea kayak, was out in the current, cutting right through the waves as my little canoe rolled up and over them.

I called Leslie and talked with her to help calm my nerves, and it got me through three or four nerve-racking miles.

We're camped at Friar's Point Boat Ramp. It's not too bad. The ramp sucked and was steep for hauling out our boats and gear, but at the top there was a large cleared out spot where we set up our tents.

It's 7:30pm. Pitch dark. And still pretty windy. If it's like this in the morning, I don't want to paddle.

We have a nice fire going.

I'm eating leftover mac and cheese from last night. Jeff isn't much of a fan of unrefrigerated leftovers, but I'm always packing them up to eat the next day.

This is our third night in a row without being able to charge our devices, so my two spare battery packs are somewhere between low and dead. I was really hoping we'll find an indoor place to stay tomorrow night. Clarksville, which is about 10 miles off the river, is our only option. It looks like it has a few lodging options. But we're not sure where we'll stop tomorrow, or how we'd get our stuff to Clarksville. I emailed one place, and Jeff left a message with a canoe company in town, so maybe they can help us out.

Take a shower. Eat a decent meal. Charge our devices.

And the Packers are on Monday night football!

Where's My Paddle, Do-Do, Do-Do

October 5, 2020 - Day 69
Friar's Point Boat Ramp to Mile 630 (Beach Camp) - 25 miles
Total Mileage: 1,699

Jeff just said something about how you plan the day, and even take into account some contingencies, like a Plan A and a Plan B, and then something takes a U-turn, and you end up having a day that you never could have imagined. From my perspective, that's life in a nutshell. It's all about what you do at the U-turn, how you handle it, that determines where you end up at the end of the day, or the end of your life.

We had another great campfire last night. Our campfires make the night way more enjoyable after camping in some not-so-great camp spots! And since its pitch dark by 7:30pm, the fires allow us to stay out of our tents for another hour or two.

We were on the water by 8:40am. Tons of big barge traffic heading up and down the river. One huge tow after another. We saw our first eight-across set of barges, each 35 feet wide, so the barge was 280 feet wide and almost a quarter mile long! We paddled two miles to some kind of barge-loading facility. Maybe soybeans?

And then there was a series of three U-shaped rock wing dams in the middle of the river. We paddled around the outside of all three to stay in the current, but still trying to keep a good distance from the barges heading upriver. After passing the three U-shaped current diverters, we got into some weird choppy, swirly water that seemed to be originating from the middle of the river and moving cross-current and upstream! Like standing waves of water that were heading to the left shore and against the current. It was weird and made me nervous. So, I angled left and tried to paddle closer to shore in case I tipped over. It seemed like I was racing these weird choppy waves, as they were always right behind me, and it seemed like I wasn't getting anywhere.

I paddled really hard for 15 to 20 minutes and slowly made my way into some calmer water, and out of nowhere, I came to a dead stop on a slightly submerged sandbar, with a bit of current still swirling around and messing with my boat.

I was a little nervous about jumping out, partly due to the weird current, as I was still several hundred yards from shore, and partly because of the quicksand we've heard about. Just a few days ago, a guy told us about a family that walked out into the river from a beach and sunk up to their hips. Two members of the family died.

I was stuck.

I carefully got out of my canoe. I didn't sink too deep into the sand. The water was only six inches deep. I grabbed my bow line and pulled my boat about 40 yards to deeper water.

Back out in the main channel of the river, the weird cross-waves had gotten worse, and a big barge was approaching, from downriver. I got back out of my boat on another submerged sandbar and waited for about ten minutes trying to figure out what to do and how long to wait. Heading back out into the main channel, and the fast current and weird squirrelly water, seemed like a dumb thing to do.

While I stood there, I started to sink down into the sand, to just below my calves. I looked back, and Jeff was several hundred yards behind me, and more in the main channel. It looked like he was stalling, waiting to see what I was going to do. It turned out that he was stuck on a sand bar, too, trying to push himself free with his paddle without having to get out of his kayak.

I was sinking deeper and decided that my only option was to get back in my canoe and paddle back out into the swirling rapids. As I climbed back in, I looked around and realized that my kayak paddle was gone!!!

Quick scan.

No paddle.

I had no idea when it had slipped out of the boat, or where the heck it was.

Now my heart was really racing, because whatever I did, I would have to do with my spare canoe paddle.

I noticed a little channel running around the back (left) side of a big sand island that was dead ahead. Choppy, swirling rapids in the main channel around the right side of the sand island, and a shallow calmer channel to the left side, that may or may not end up in a dead end.

I opted for the left channel, hopped in my boat, and paddled like my life depended on it against the current that was trying to drag me into the main channel. A few minutes later, I was in calmer, more protected water.

I had enough cell service to call Jeff, who had gotten lose from his sandbar, and was heading through the main channel. It turns out that his rudder was tuck, and he had poor maneuverability, but he eventually pulled up on the outside of the sand island. He agreed to walk across and handed me his extra kayak paddle.

I felt stupid for having lost my main paddle, but thankful that we had spares. Paddling a solo canoe in this kind of water with a canoe paddle, just doesn't provide near as much control and power as using a two-bladed kayak paddle. Jeff is nervous about paddling without his spare now.

By 11am, we were back to just paddling down the river. It took over two hours to paddle the first three miles of the day. Not a great start. I was still a little rattled, and happy to just be moving down the river, getting used to Jeff's shorter, heavier, loaner kayak paddle. I was happy to be out of danger, and downriver from all the excitement.

For the rest of the day, I just had a profound sense of appreciation for the slow-moving current, fewer barges, and my new paddle. Appreciative of non-epic paddling. Jeff later told me that he took a wave in his chest when he paddled to the right side of that sand island. I'm so glad I didn't go that way with my canoe paddle.

I was able to call John at Quaypaw Canoe Company to see if he had a spare kayak paddle that he was willing to sell, and someplace where he could meet up with us later in the day tomorrow. John said that he could get a driver to bring me a couple of paddle options, but the driver could only meet us at boat ramps that were still 20 and 35 miles downriver. Since we took a couple of hours off of our paddling day, due to my lost paddle debacle, we ended up camping for the night around mile 25. We have no cell signal here, so no internet, and no phone. We'll have to try and call him again downriver sometime tomorrow.

We'd originally gotten John's number from an Airbnb in Clarksdale. I was hoping we'd be at that Airbnb tonight. But there just isn't much river access by road along this stretch of the river. We asked John if we could pay someone on his staff to meet us somewhere along the river today and bring us to Clarksdale tonight. We're both ready for a shower, hotel, and electricity. Both of my back-up battery packs are dead. Jeff loaned me one of his three today so that I could charge my phone.

One of the reasons we stopped at the 25-mile point today was to set up our solar chargers and attempt to charge something up. So, at 25 miles, we spotted a nice sandy campsite with good sun exposure to the southwest, stopping around 3:30pm.

I made dinner tonight . . . instant potatoes with bacon bits, and a separate pot of ramen noodles with pizza sauce. Lame but filling. Have I already written that it's time for a motel and restaurant?

It's 6pm. Dishes are done. Since I cooked and washed dishes down by the river, Jeff is collecting wood for tonight's fire. He's a good fire builder. It's going to get down to the mid-40s again tonight. Great sleeping weather. Time to get out my book since there is no cell/internet service.

Seth is flying in Texas; Tyler and Dana are at our cabin in Michigan; Leslie in home in Indiana; Arlo and Shauna are enjoying each other; and I'm camping along the Mississippi River with my buddy Jeff, focusing on really appreciating where I am right now.

Finally, A Nice Break in the Action

October 6, 2020 - Day 70
Mile 630 (Beach Camp) to Dennis Landing Boat Ramp - 20 miles
Total Mileage: 1,719

I'm not going to write much tonight. It's already 10:20pm. We are in a nice Hampton Inn in Cleveland, Mississippi. Everything went as planned.

We woke to a chilly and dewy morning. Coffee and Captain Crunch with powdered milk were on the breakfast menu. We were on the water by 8:30am. A little fog hung over the river. Cool paddling temps. A little current helped us along, and we only passed two big barge tows all day.

We averaged six miles per hour for the first two hours, which was our fastest yet.

We pulled out at 20 miles like we'd planned at a totally destroyed and no longer used boat ramp. The parking area entrance was blocked off since the ramp is unusable. So, we didn't feel too worried about leaving our boats and gear for a day. We had cell service where we pulled off, which was awesome.

We called a taxi in Cleveland which is about 35 miles away. The guy had absolutely no idea where this boat ramp is. It's literally in the middle of nowhere. It took three calls to direct him to where we were sitting with our bags on the side of a graveled levee road. Not a single car passed during the 90 minutes we waited for the taxi to show up.

Hotel. Showers. Laundry. And an awesome, very late lunch. I did a few hours of work on my computer in the lobby.

John the canoe outfitter guy agreed to send his lead river guide, Mike River (literal last name), over at 7:30am to bring me a few paddles . . . one of which I'll buy for a spare. Mike will then drive us back to our boats, hopefully after a quick stop at McDonald's.

Jeff and I are at odds tonight. I'll write more about it the next few days as things play out. But I'm getting the feeling he may be giving up on this trip. He talked about Chris coming to get us tomorrow, or the next day!! WTF?

We had a very direct and confrontational discussion tonight. He seems to be back on track, but I'm not 100 percent convinced. We'll see.

With about 18 days to go to get to the gulf, I do not want to finish this trip alone. The issue a combination of some personal matters that Jeff needs to deal with at home, and a hurricane that is hitting the gulf this Friday and Saturday, which will for sure push us off the river for at least a day or two, or maybe three.

Jeff just said, "Everything is falling apart. I won't blame you if you decide to bag it."

What?

Me? Bag it?

Nope.

We're going to finish what we started.

Like always, let's just take this one day at a time. Just like we always do.

Tomorrow we need to paddle 30 to 33 miles to be on track to get to Greenville by Thursday. So, let's just focus on that, Jeff. We'll just focus on that.

Take the Long Way Home

October 7, 2020 - Day 71
Dennis Landing Boat Ramp to River Mile 585 - 25 miles
Total Mileage: 1,744

Sitting by our nightly campfire on a flat spot above a sandy beach. There are at least eight lit-up tugboats tied up or moving barges around on the other side of the river. We are in the middle of nowhere but camped across the river from a barge terminal that has six or seven companies with their own loading docks. So, there is a lot of barge activity. The tugs look cool all lit up after dark. At least a dozen big barge tows passed us today. And the channel was pretty narrow in a few places, so we actually had to paddle pretty close to a couple of them.

When we shoved off this morning, a huge barge was coming right towards our boat launch site and probably only passed 50 yards away. I took a photo of Jeff having just gotten into his boat, and this huge barge passing very close by.

We didn't start until 11am today! Mark River from Quaypaw Outfitters agreed to give us a ride from the hotel back to where we'd left our boats at Dennis Landing. Mark was super nice. He's the head river guide for the company, so he knows all about this section of the river. But he charged us $125 for the ride, compared to the $70 we paid for a taxi yesterday afternoon, and he took the most round-about way. It was weird. I was following our route on Google maps on my phone, and Mark drove an unnecessary 50 extra miles more than the taxi guy.

Consequently, he picked us up around 8am and we weren't to our boats until after 10am, for what, yesterday, was a 35-mile drive. Two hours! Oh well.

A late start for sure. And the river was super slow today. The first 10 miles took almost three hours to paddle. We just didn't seem to get anywhere. The temps were in the low 80s with not a cloud in the sky. Several times I dunked my Packer hat in the river and put it on my head, just to let the cold water run down my back. Or I'd dip my coffee mug in the river and pour it over my head to cool off.

Speaking of which . . . my boat must have a few small leaks or cracks. By the end of each day, I probably have a gallon of water sloshing around in the stern of my boat, which is the heavy end. I put most of the weight behind me and towards the back of the boat, so the water is collecting back there. I have a sponge that I use to sop up water that starts creeping forward. I'm sure that dumping cups of water on my head throughout the day added to the water that is seeping in.

I did end up buying a kayak paddle from John Roskey. He sent five different ones along with Mike this morning. A couple of cheapies, and a couple of very nice ones. I chose the longest nice one, and it still ended up being way too short for me. But I'll make it work. He charged me $285 for a used paddle. But I know it's a great paddle. Werner is the brand. It's probably a $400 paddle new. I also ordered a new paddle yesterday and am having it sent to a guy named Evan in the town of Delta, Louisiana. So, it should be there when we get to Vicksburg (which is across the river from Delta), in a week or so. The one I ordered is longer.

Sunny day. Lots of boats. I listened to the radio a lot. I kept telling myself how lucky and appreciative I am to be paddling on the Mississippi River. It's so easy to wish the days away and not appreciate them for what a great and unique experience this is.

I had to give Jeff a stern talking to last night about not giving up and giving in . . . that we are in this adventure together and we're going to finish this damn river one way or another. Leslie won't let me come home until I get to the end. I know that she's ready for me to move on to a new obsession. A new adventure. She said as much the other day. This one has been the primary topic of conversation for a couple of years now.

I'm always doing the math of how much further we have to go—how many miles, how many days. I think we're at 17 more paddling days at an average of 25 miles per day, which is super doable. There is a hurricane coming, so that'll slow us down a day or two. And then Jeff needs to drive back to Nashville next week. Uggh. Don't get me started on that. It's why we had a tense stand-off last night and a lot of quiet time together this morning.

Mark told us that we'll definitely see alligators and water moccasins on the Atchafalaya River. That'll be wild.

Tomorrow we are being met by a River Angel, which means a warm, dry place to sleep, and hopefully a shower and some shopping. I need to buy some Advil, bug spray, another power charging brick, a couple of dinners, bagels, cold cereal, and some powdered milk. We should get our miles in tomorrow and Friday before the hurricane winds and rain make it this far up the river.

Park Neff, The Man, The Legend

October 8, 2020 - Day 72
River Mile 585 to River Mile 562 - 23 miles
Total Mileage: 1,767

Our wait for Hurricane Delta begins.

A massive Category 4 hurricane is scheduled to slam the Louisiana coast and then barrel straight up the Mississippi River early tomorrow morning. But the front edge has already started to hit where we are. It was overcast all day and started to rain just as we pulled off the river around 2pm today. It's been drizzling on and off ever since.

I had really hoped we could paddle tomorrow (Friday). The forecast is for 10 to 15mph winds and rain, building to heavy rain and higher winds by evening.

Park Neff, the River Angel who met us at the boat ramp this afternoon, and whose cottage we are staying in tonight, said that we shouldn't even think about paddling tomorrow. And I trust his judgment. He knows this river, and he knows hurricanes.

So, we will definitely NOT be paddling tomorrow, or Saturday for that matter. It bums me out to add two full days to the trip, but we are safe and dry, and have a great place to "hunker down," as Park Neff has said a couple of times already.

The paddling was pretty uneventful today. Only 23 miles to the boat ramp. It was either stop where we did, or go another 17 miles, which we didn't want to do. We had to get to a place where Park could pick us up. And we needed Park, because a hurricane is heading our way. We passed only two big barge tows today. Likely, because they are hunkering down too.

Very little current again! The unusually low water level this fall has just slowed the river way down. So, even paddling 23 miles was work.

Park Neff is a big-time paddler. He and his wife have been hosting river paddlers for years. He's a river paddler, and a Mississippi River through-paddler himself. Park has also paddled with several well-known cross-country, cross-Canada paddlers who have done all kinds of epic and crazy adventures. Tons of stories and very entertaining.

Park also knows several River Angels that live further downriver that we can connect with. He is also a big fan of finishing the Mississippi River by crossing over to the Atchafalaya River that drains into the gulf. So, that decision that we've been contemplating for months is a done deal. Park is also suggesting a variation for the last 30 miles of the Atchafalaya that sounds super interesting that cuts off a day of paddling and has a road at the end where we can be picked up.

After we got set up in their guest house, Park and Sharon took us out to dinner for pizza. Park is a Baptist minister, and a farmer. When we got to his truck, Jeff said something about wearing COVID masks, and Park said, "I'm an anti-masker." Okay. A Baptist minister in the Deep South who is an anti-masker. Who would have guessed?

When we returned from dinner, Park projected his computer screen onto his big-screen TV and showed us the details of this alternate route he's suggesting to get to the gulf. It's basically cutting over to the Atchafalaya River, then to Yellow Bayou, and then to Hog Bayou, and on into the gulf. Whenever his computer cut out, Tucker Carlson from Fox News came onto the TV.

Note to self: Refrain from swearing, making any political comments, or talking about religion. Let's just keep things amicable. We are going to be here three or more days while we wait out Hurricane Delta, and Park and his wife are the most amazing hosts imaginable.

Hurricane Delta

October 9, 2020 (Day 73) to October 10, 2020 (Day 74)
Park Neff's Cottage Waiting Out Hurricane Delta
Zero Days

We spent two days at Park and Sharon's paddler's cottage. For several hours both days I caught up on computer work. We also hung out with Park a bit and treated them to dinner one evening.

I decided that I have to fix my canoe rudder, which is completely inoperable. It's been getting stiffer and harder to use for the past few weeks, and simply stopped working a few days ago. Paddling a solo canoe on the Mississippi River makes it pretty much a necessity to have a rudder, especially when the wind and waves get intense. So, I got online yesterday morning and found a kayak store in California who had a rudder system replacement kit. They agreed to overnight the kit, and it amazingly arrived today, on a Saturday.

I had to completely dissemble the right pedal system. I asked Jeff to come out and help me remember how things were coming apart so I could put them back together again. And then Park came out, and he was super helpful. We ended up having to drill two holes in the metal gunnel so I could feed the spare cable along to connect the pedal to the rudder. It seems to be working better. We'll see once I'm back on the water.

We're both very anxious to start paddling on Sunday.

Ups and Downs and Ups and Downs

October 11, 2020 - Day 75
Mile 52 to American Bar Island - 36 miles
Total Mileage: 1,803

Today was a little up and down for me. Thankfully, it ended on the up and not the down. A little Redemption whiskey and sitting by a nice fire. Probably our nicest campsite yet. We're on a big sandbar, about 100 feet from the river, and maybe 10 feet above the water line. We should be okay. Park thinks that the river may rise a foot or so overnight from storm surge. The mosquitoes are bad. That's the only bad thing. After the hurricane and more than seven inches of rain, the skeeters are out.

Our boats were extra heavy starting out this morning. I have two and a half gallons of water, and probably way too much food. All three of my food bags are heavy and overstuffed. I'm not sure what I'm saving it all for? I know that I have enough food for these final two weeks. I told Jeff not to let me buy any more food!

I've been excited these last two days, sitting out this hurricane. Anxious to get moving again. Eager to get some miles under our belts. We have about 14 days and counting. I charted out the remaining miles, about 300 to go, and I have us finishing two weeks from today. So, 15 total days. I was all excited to get moving, until I actually started paddling at 8:15am.

After only a few miles, I remembered that even though we only have 15 days to go, it's still 15 days of paddling all day long.

After just a couple of hours of paddling, I remembered that paddling all day long is a shit load of paddling, and not that much fun to do hour after hour after hour. We had a decent current pretty much all day today. It took us eight hours to go 36 miles, so 4.5mph that included a 45-minute stop at the boat ramp in Greenville.

But by 10am, I tried calling Leslie, Seth, Aaron, and my mom, just for a distraction from paddling. My mom was the only one that answered initially. I just needed to talk, to get my mind off paddling, and to help me pass the time. It definitely helped. I eventually ended up talking quite a while with Aaron, and had a short conversation with Seth, and then Leslie.

Our Plan A was to stop at the Greenville boat ramp at mile 25, but we were there by 1:15pm. I needed to get out and stretch. And to try to readjust my rudder cable. There was too much slack in my right cable, and I had a hard time turning my canoe to the right. When the water is moving fairly fast like it was today, I need my rudder more than usual to control my boat.

I was reminded what a big river this is.

The water moved right along, and when a breeze from the southwest started picking up and causing waves and white caps, I became reacquainted with my fear of the water.

We pulled off at the boat ramp to stretch, poop, and eat something for lunch. The ramp is right next to a big channel that leads to several big unloading docks. And the barges coming upstream and heading into that channel came really close to the ramp. And, since they were heading upstream, they kicked up ten-foot-high rooster tail waves that ricocheted back and forth in the channel for a full 15 to 20 minutes after the tug has passed.

It wasn't a very restful stop because we had to rush down to stabilize the boats when big waves crashed ashore from the passing barges. And the banks were super muddy, so I sunk down a foot or so into the mud while trying to keep the waves out of my beached canoe.

I was getting progressively more nervous about pulling back out into the main river channel after I'd readjusted my rudder and considered just stopping for the day. The side channel emptied into the main river right where the river turned to the left, so we couldn't see what was coming upriver just around the corner.

The big waves made me feel anxious.

So, we waited for a second, smaller barge to pass, and then ventured out. I told Jeff that if the waves got too big, I was going to run around and head back to the ramp. But, once we got out around the rocky corner, the water calmed down quite a bit.

I love the adventure of this trip—the challenge, the motivation and commitment that it requires. But I do not like the big water, waves, wind, and massive tow barges.

Not at all!

We decided to push on 10 to 11 more miles to American Bar, a huge sand island that Park had recommended as a camp spot.

The last 10 miles were great. The river was super calm. Almost perfectly flat. We had a little current, and a slight breeze at our backs.

Overcast. Maybe high 70s.

It was perfect paddling conditions and we both remarked, maybe the best of the entire trip. We're camped on an awesome sandy beach and have a nice beach fire going. One of our nicest campsites for sure, except for the dumb mosquitoes.

We can tell that the river is a little higher than it was three days ago, before the hurricane. Higher and faster for sure. Being at the end of the day feels great. We went further than we'd planned and have a beautiful place to enjoy the sunset and evening.

Tomorrow, there is supposed to be wind from the south, so that will slow us down a bit. We're hoping to go 31 miles to the tiny town of Mayersville, Mississippi. There is a small grocery store there, so something to look forward to.

Our upcoming goals: Vicksburg, Natchez, Lock and Dam to the Atchafalaya, Atchafalaya River, Hog Bijou, Gulf of Mexico.

Park did a little witnessing as he drove us the 45 minutes back to the boat ramp this morning. We heard how he went from being a racist to giving his life to Christ. I knew he'd try to squeeze it in somewhere during our stay. Baptist Minister. What do you expect?

Big Head Todd and the Piglets
October 12, 2020 - Day 76
American Bar Island to Mile 515 - 12 miles
Total Mileage: 1,815

Well.

Not a lot to show for our efforts today.

Twelve measly miles.

We'd hoped to go 31 to Mayersville, which has Tony's Grocery.

We were up and out of the tents by 7am, and on the water by 8:05am. A record for Jeff getting his shit together.

But we knew that there was going to be a wind out of the southwest, and there was. Initially blowing 10 to 15mph and building to 15 to 20mph. It made for slow progress, and lots of turbulence and waves. I was fighting waves all morning. Never really scary, but always tense. Despite my rudder repair job, it still isn't working right. I paddled tense the whole time.

There was a decent current, so when paddling hard we could go about three miles per hour, but the wind blowing against the current made a lot of waves. So, we stayed pretty close to shore, which tends to make the waves even worse as the water pushes up on to shallower water. More eddies, and backflowing current, and more squirrelly water racing around the wing dams.

After about two hours, I pulled off just to regroup a bit. Just before I'd stopped paddling, and was digging around in my bag for sunscreen, I took a wave over the side of my boat.

So, fuck it. I pulled over and was ready to stop for the day right then and there. If Jeff had gotten out of his kayak, I would've tried convincing him to stop. But he just floated around and waited for me to finish my short break.

As the wind and waves got worse, it got to the point where we were paddling hard and barely moving, and about to turn a corner into even more of a head-on wind. I spotted a nice sandy beach and some trees offering shade. It was around noon. I wanted to take a break, so we stopped for lunch and just sat around for an hour or two to see what happened with the wind.

For an hour we hemmed and hawed about the pros and cons of getting back into our boats and pushing on into the worsening wind, we decided to just stay where we were. Plus, we were already at a great camping spot. Who knows what we'd find up ahead?

We have about 80 miles to Vicksburg. Two 30-mile days and then a 20 on Thursday. So, paddling another few miles today, into the wind, wouldn't make much of a difference.

Neither of us really like just sitting around, but I hate paddling into a stiff wind and waves and barely getting anywhere. Staying here, at least we know the consequences. Paddling on . . . who knows?

It got hot today, like mid-80s, but was cloudy and breezy for the most part. We're camped on sand by some trees. It's definitely a hunting or fishing camp of some sort. There are two flat-bottomed boats on trailers parked up in the weeds, a big fire pit, a plastic folding table, a big charcoal smoker/grill, and an American flag tied to a tree and proudly waving.

So, we set up camp. Sat around. Took a short nap. And then decided to take a walk about a mile to the levee road just to stretch our legs. The mosquitoes were horrible back in the woods and away from the river breeze. One swat would kill three or four mosquitoes every time. Satisfying, but also depressing.

We got to the levee just as a tractor that was pulling a road grader pulled up.

Todd was friendly and curious about why two men were walking along his levee road, miles from anywhere. Todd owns the camp where we'd pitched our tents and apparently the island directly on the other side of the river. He said that there was a small Roy's grocery store three miles away and offered to come back to get us when he's done grading and bring us to the store for cold beer. So, rather than starting to cook dinner, we went back to camp and depending on when he comes back to pick us up, maybe we'll just get something for dinner at Roy's.

Well . . . what an evening!

There could be an entire 12-episode reality TV show about what happened next. I wish I could write it all down.

So much fun. So hilarious. Definitely a highlight of our entire trip.

Todd and his wife Paula returned in their four-seater side-by-side to pick us up. Cold beers in hand for each of us. We drove to their house that was about a mile and a half away, with Todd pointing out various things along the way.

We pulled up to an old farmhouse and huge garage/barn with what looked like everything they owned sitting out in the yard. A farmer hoarder. There were even ten little two-month-old piglets running around in the yard. Paula invited us to stay for dinner—red beans and rice. Park Neff told us that if anyone in the Deep South invites you over for red beans and rice, get ready for a feast!

Todd drove us four miles in the four-wheeler to Roy's grocery store, a little store that had seen better days and that was over 100 years old. There were only a couple of shelves of food, but there were four people sitting inside at a counter, each with a bottle of

Bud Light. We got to talking about our Mississippi River trip. They were all pretty impressed.

Back at the farm, Todd drove us out to check on his 200 cows. Todd clearly loves his cows. He has named every one of them. Names like Bucket Head, Lu 342, and Zebra. We got out of the truck and Todd petted the cows that gathered around us as we were told about that particular cow's personality and quirks. After meeting several cows, Todd drove us all over his property and told us about his soybeans, wheat, and corn. Todd has lived on the Mississippi River for his entire life.

When we got back to the house, Todd fired up his smoker, and smoked the most amazing pineapple infused sausages, and a few dozen chicken thighs. He made the sausages himself. Meanwhile, Paula was whipping up an amazing southern dinner herself. No one was in a hurry.

Todd also fired up a huge deep-fryer and while we drank ice-cold Bud Lights, one after another after another, and listened to Todd's crazy stories about other river paddlers he's dragged home over the years, Todd started deep frying Mexican okra (jalapeños), corn balls that were delicious, and finally deep-fried macaroni and cheese.

Throughout the evening, Todd told us about dead bodies that have washed up on the riverbank, neighbors who turned out to be convicted pedophiles who were on the run from the police, a river traveler who was down on his luck and living with cancer who had a hole in his boat, so Todd gave him one of his boats. The stories, and the cold beer, and the food just kept on rolling out.

And just when our stomachs were so full they were about to bust, Paula brought out the most amazing red beans and rice I've ever had with homemade cornbread. I don't even remember what the desert was, due to my food coma and alcohol haze, but I do remember when Todd's nephew stopped by and whipped out two bottles of homemade moonshine! One even had a skull and crossbones drawn on the label.

Todd and Paula drove us back to our tents sometime late in the evening, loaded down with leftover sausage, chicken, cornbread, and a gallon of homemade sweet tea. And then about 15 minutes later, they returned with plastic cups for the sweet tea and a bag of tortilla chips that they were sure we'd need on our paddle tomorrow.

What an amazing night.

What amazing, giving people.

All made possible because we quit paddling after going only 12 miles!!

Jeff's Near-Death Encounter with a Huge Tow

October 13, 2020 - Day 77
Mile 515 to Mile 480 - 35 miles
Total Mileage: 1,850

It was a good paddling day today. Paddling 35 miles, even in good conditions is a hell-of-a-long way to go. And sitting eight hours in the boat is just a long damn time. Too long really. Ideally, for me, a three- to four-hour paddle every day would be perfect. I think Jeff enjoys sitting in his kayak more than I enjoy sitting in my canoe.

We were on the water at 8:15am. It took a while to find spots to stow all of the extra food and sweet tea that we'd been gifted from Paula and Todd.

It was a cool morning. Slight breeze out of the north. We enjoyed a bit of current for most of the day too. There were a few parts of the river today where we felt like we were paddling through molasses. There was another four- to five-mile section where we had a strong wind blowing in our faces. But for the most part, we moved along, and the paddling conditions were way better than yesterday.

It was cloudless, so even though the temps only got to the mid-70s, I felt like I was roasting in the bright sun.

There were lots of big barges again today, too, including a few 35s, 42s, and even a 56-barge tow!! It's hard to describe how big that is, especially when you are passing by in a 13-foot canoe.

The excitement of the day came around mile 30. The river was moving especially fast, which I've already explained makes me very nervous. The wind had also picked up to a steady 15mph. We started paddling past a series of weird, standalone rock dikes that were sitting out in the middle of the channel, not connected to shore like the typical rock dams that we'd passed a hundred times. These rock dikes ran completely parallel to the river and served to funnel water into the middle of the channel and really speed up the current.

So, we were whipping along, faster than any previous portion of the river up until now. And a big 42-barge tow passed us coming upriver. The first long standalone rock dam was about 500 feet long and was coming up fast. We each had to decide whether to cut to the left and duck in behind the rock dam or stay out in the main channel to the right. Jeff was too far ahead to hear me yell that I was going to cut left and out of the main channel. It was a split-second decision as I noticed another huge 42-barge tow heading upstream and took some serious paddling with my non-functioning rudder to steer to the left. But it just seemed a little calmer and safer to head out of the main channel.

By heading to the left of the long rock dam, I was hoping for a little protection from the inevitable oncoming wave wash from the massive tug engines that would be coming our way as the barge passed. So, I cut to the left and inside the rock wall, and saw Jeff first try to cut to the left as well, but then he must've realized that the water was moving too fast, and he didn't have enough time to go left, so he swung to the right, and headed outside the rock dam and was quickly swept along by the main current, which had really picked up since these rock dams were doing what they were supposed to be doing.

As the current whipped me along the left shore, I lost sight of Jeff paddling on the other side of the long rock dam. What I could see was this huge set of 42 barges, stretching over 1,000 feet long, heading upriver and passing really close to the rock wall. From my vantage point, it looked like the barges were almost scraping along the rocks they were so close, and I was terrified to realize that Jeff and his tiny kayak were somehow between the barges and the rock wall.

The current on my side of the wall was still very fast and choppy, so I had to pay attention to myself to keep from tipping over.

I eventually was able to get closer to shore and in calm enough water that I could turn my canoe slightly and look back to watch as the huge barge came within feet of the side of the rock dam, with no sign of Jeff. For a moment, I hoped that maybe he'd turned and tried paddling back upstream to get back around the top end of the rock dam to avoid the huge barge. He told me later that he'd tried, but the current was too swift.

So, I watched for a few tense minutes as the barge passed the rock dam. I started to think about what I was going to tell Jeff's wife Chris. I knew that Jeff was in a very dangerous situation, and that his kayak had likely already flipped over, and he was swimming within feet of the huge tug propellers. I hoped that he would be able to slide out of his kayak and swim over and grab onto the rocks.

Finally, after what seemed like a very long time, I saw Jeff's tiny kayak, popping out of the downriver end of the rock wall, and right alongside the last of the cabled together barges. And he was paddling his ass off, in my direction, trying to get away from it all. In retelling the story later, he said that he kept getting sucked back towards the moving barges. He was also getting close to the "wheel wash" from the tug that was causing 10-foot-tall waves.

I was still getting swept along quickly and had to pay attention to where the current was taking me, so only was able to catch glimpses of Jeff back over my right shoulder. Jeff was eventually able to paddle out and headed over to where I was, trying to stay in place and watch how this was all going to play out.

When Jeff got to me, he was scared shitless. That was a very, very close call, and he knew it. He said later that there were a few minutes where he thought he was going to crash into the rocks because he had to paddle so close, and then flip and get sucked underneath the passing barges.

It was a close call.

It really shook both of us up.

At mile 30 or so, we were done and started looking for a spot to pull off but didn't find anything until mile 35.

A long day.

We have another great campfire blazing.

Dinner of leftover smoked sausages, mixed in with noodles and pizza sauce. It was a great dinner. I also had a couple of leftover pieces of unrefrigerated smoked chicken that Jeff wouldn't touch because they'd been in the heat all day.

Good day. Long day. Hard day. Sunny day.

Forty-three miles to Vicksburg. We're both looking forward to getting there on Thursday. I had a nice long phone call with Seth tonight. And a nice long texting conversation with Leslie.

I'm glad the day is over. And I'm glad we are both safe!!

Thank God for Dragonflies

October 14, 2020 - Day 78
Mile 480 to Mile 453 - 27 miles
Total Mileage: 1,877

Not much to write about today. I guess that's good, because yesterday I was writing about Jeff's near-death. No near-death experiences today. Just paddling. Actually, very few barges passed us today. Maybe four? Which is a nice change.

It was chilly last night. We had a really nice fire, some bourbon, and turned in around 9pm. I read a few pages in my book before falling asleep. I'm not sleeping very soundly on this trip. I wake up at least six or eight times a night to either roll over to my other side because the side I've been sleeping on is achy and sore, or I have to pee and need to roll over to get the right angle into my Nalgene pee bottle. Plus, we are in our tents for nine to 10 hours a night due to the shortening days, so I'm just not tired enough to sleep that long.

There was a heavy dew again this morning. The tent fly was dripping wet. I'm never a fan of packing up a soaking wet tent. It was also a hassle unloading and reloading our boats at this site because we were camped up on a bluff. But the view out over the river was beautiful, and we had a nice breeze up here that helped keep the mosquitoes at bay.

Today we just paddled. Neither of us had too much energy or motivation. We'd planned on paddling 30 miles but checked out an alleged boat ramp at mile 23. If it had been a nice park-like spot, we'd have stopped. But we couldn't even find it as we passed the location that was marked on our Army Corps maps. So, we paddled on with the mantra that the more we go today, the less we have to paddle into Vicksburg tomorrow, and the more it'll seem like a day off.

By mile 26 or 27 we were done. So, we pulled up on a beautiful sandy beach. The sun was hot, and the sky cloudless. We lugged our packs several hundred yards so that we could set up in some trees and get out of the sun.

I had another long phone call with Seth while I carried my stuff from the boat to the site, and while I set up my tent.

The sun was really beating me down, especially later in the day. We set our camp chairs up in the shade and split a warm 16-ounce Bud Light. It was sort of a metaphor for the day. Just sort of blah. I kept trying to get myself fired up by saying, "I'm on the Mississippi River. This is an amazing adventure." But it didn't work today.

Mosquitoes are horrible!

We bought some citronella spray that only lasts an hour or so. So, bring on the 100% DEET! Skin cancer be damned!

The sun is setting and it's only 6:30pm. I collected firewood for a small fire tonight. We're both tired and want to hit the river a little earlier tomorrow morning so we'll probably be in our tents by 8pm.I'm looking forward to taking a full day off in Vicksburg while Jeff drives back to Nashville to close on his condo purchase. I have work piling up, laundry to do, and can't wait to take a shower. I've been sticky for days. Sand is everywhere, and in everything. I want to go through my three food bags too and see what I've been lugging around for the past 32 days. Might be time to get rid of a few things.

We only have nine more paddling days left after tomorrow.

Hopefully.

I can't keep putting the rest of my life on hold for much longer.

Beautiful sunset. Dragonflies are dive bombing the mosquitoes.

Nice!

Paddling Into Vicksburg

October 15, 2020 - Day 79
Mile 453 to Vicksburg, Mississippi - 17 miles
Total Mileage: 1,894

I woke up at 6:45am to the sound of Jeff walking past my tent, hauling some of his dry bags to the river in the pitch dark. I hadn't heard him get up and out of his tent.

We ended up turning this spot into a nice campsite and had another great fire last night. And we both had a little Redemption (my new favorite bourbon). But our site is a long fucking walk from the river. It took seven or eight minutes to haul a load of gear to the boats, and then walk back. We hauled our boats about half of the way, just to get them high enough above the water line in case the river rose overnight. We usually bring our boats way too far above the water line, but can you imagine waking up and seeing your boat gone?

There was a thick fog hanging over the water this morning. When I saw Jeff outside my tent this morning at 6:45am, he still had his headlamp on, and the sun hadn't yet come up over the horizon. Jeff has been saying that the barge tows don't travel in the fog for safety reasons. I never bought that line and couldn't help pointing out the two huge barges that were coming towards us as we first started paddling at 8am.

It was a cool morning. Thank Yahweh! Cool in terms of beautiful and calm and temperature. And different from any other morning so far. I loved it!

Boats appeared out of nowhere. The rising sun struggled to peer through the fog. I took lots of what turned out to be great photos.

We "only" had to paddle 17 miles today, but that's still a four- to five-hour non-stop paddle. I had my big bowl of oatmeal before I headed out, and a nice hot insulated mug of instant Starbucks at my feet.

There was more of a breeze this morning than predicted on our weather apps. We both obsess about the weather throughout the day. The wind blew maybe 8mph into our faces for at least the first 10 miles. A little disheartening, but not horrible.

We crossed the massive river early on to cut a corner, so then had to paddle back across to be on the correct side to exit at Vicksburg. I really don't like paddling across the middle of the river when it's so huge here on the south end of the Mississippi. It's just too damn big. We try to pick spots that look narrower on the map, but there is still current, and today there was wind, and always the possibility of massive barges coming in either direction. It's amazing how a 1,000-foot-long barge can "sneak up" on us from behind sometimes. So, my heart is always in my throat when we are crossing from one side to the other.

We could see Vicksburg off in the distance when we were still five miles away. It seemed like we'd never get there. But the fog eventually burned off, the breeze lessened, and it was some very enjoyable paddling.

The city boat ramp for Vicksburg lies up a mile-and-a-half-long narrow channel. The second we turned left into the Yazoo Chute, we were hit with a 4-5mph current coming towards us as this river channel drained into the big river.

It was horrible!

We paddled as hard as we could, literally, and moved only inches at a time. After 15 minutes in the channel, I looked over at Jeff and he'd moved maybe 15 to 20 yards. Twice I turned my boat around and thought, "Fuck it," knowing it could take hours to travel a mile and a half in this strong current that was pushing against us. The maps didn't show any place to pull off further downriver, but I was willing to try to find something.

There were lots of tugs tied off along both sides of the channel. At one point, we were struggling past a tied-off tug, and I asked the crew, "How far to the public ramp?" One guy said a mile, and the other guy a half-mile. Grrrr. It was non-stop paddling as hard as I could just to keep from going backwards.

Staying just a few feet from the very muddy shore helped a little. As we passed very close to the side of a tied off barge, a guy yelled down, "Stay away from the bow of my boat so you don't get sucked under, in the undertow."

Yikes.

We eventually made it to the dead carp strewn ramp. Stinky!! But we were so glad to be there. The ramp is right in the center of town. It ended up taking about 90 minutes to paddle what was only one mile.

A River Angel that we'd contacted in advance, Layne, showed up about 45 minutes later, and drove us to the Duff Green Mansion where I'd made a reservation for two nights. And then Layne took Jeff to pick up his rental car.

Duff Green Mansion is very cool. Very fancy. We put our boats in the fenced backyard by the drained swimming pool and hauled all of our stuff to the room. And then Jeff jumped in his rental car and headed to Nashville.

I unpacked, showered, put on some clean clothes, and headed downtown to historic Washington Street to a great micro-brewery. I had some amazing chicken wings and a couple of good IPAs as I looked through my newly arrived Atchafalaya River maps that Leslie mailed to our contact, Evan, in Delta, Alabama, just across the river from Vicksburg. Earlier today, Evan had dropped off my new maps, and my new replacement kayak paddle that I'd also mailed to him. Christmas in Vicksburg.

Tonight, I'm watching the news. Nineteen days to the Presidential election. Please, dear God, send Trump packing!

Tomorrow my plan is to really focus on getting some work done, as well as drying out my tent, going through my food, washing my and Jeff's clothes, washing my nasty sandy dishes, and mailing home my now extra kayak paddle. Jeff should be back by 8pm tomorrow night, so I'll have the whole day to get stuff done. And then it's back on the river. Hopefully nine more paddling days to go. Maybe 10.

October 16, 2020, Day 80
Vicksburg, Mississippi - 0 miles
Total Mileage: 1,894

Relaxing day.

Breakfast at the Duff Green Mansion is a fancy affair. Guest are invited to a very formal dining room that is full of antiques and old paintings. Following a brief history of the role that the mansion played during the Civil War, a multi-course breakfast of fruit, rolls, eggs, and champagne were brought out by a host and servers dressed in period costumes. The other overnight guests were mostly well-dressed older couples staying in the area to tour the nearby Vicksburg battlefields. I skipped the guided tour of the mansion that was offered after breakfast by a woman wearing a frilly 18th century gown. Not my kind of thing.

After breakfast, I walked down the hill to Washington Street and bought a good cup of coffee. I did a quick walk-through of the Lower Mississippi River Museum. I'm not much into museums and actually went inside because I had to poop. There were several interpretive displays about the Army Corps of Engineers, and the re-routing of the river over the years with huge locks and dams and dredging over the decades.

After wandering around a bit, and mailing home my extra paddle, I discovered Mama's Super Burger, "Where You Taste the Meat Before the Bun." It was a tiny little one-room place in a less affluent part of town that was run by an older black woman who was as nice as could be. There were two painted picnic tables out front, and you place your order through a sliding window. I got a huge hamburger with everything on it, a huge order of nachos with ground beef and jalapeños, and a large Coke, all for $8.00!! It was so much food that I couldn't finish it all. I was the only one there, and glad I'd stopped.

Back at Duff Green Mansion, I got a bunch of work done, put out a few fires, and caught up on emails. I tried to take a nap at one point today but couldn't fall asleep.

Jeff was back by 5pm, having completed his real estate transaction in Nashville. We walked down to the same place I ate at last night for beer and dinner. We sat up on the rooftop, next to a faux fire pit that burned gas, and splurged on some good food, good beer, and a couple of after-dinner whiskies. It was good to have Jeff back. It's such a weird and disorienting experience to be on the mighty

Mississippi River fighting wind and waves and feeling intense fear on one day and sitting at a rooftop restaurant sipping whiskey by a fake fire the next day while contemplating getting back on the river.

Back in our room, I spent some time organizing my food and gear, and restuffed my now dry tent. We're all ready to head out first thing in the morning.

I'm feeling ambivalent about getting started again. But once we're back on the river it'll be fine. Our goal is to go 30 miles tomorrow. We can do that. Just stay focused on going 30 miles every day, and we'll be done in nine days!!

Peace and Solitude Next to a Nuclear Power Plant

October 17, 2020 - Day 81
Vicksburg, Mississippi, to Mile 408 - 31 miles
Total Mileage: 1,925

Not a lot to write about today.

Ricky, the co-owner of the mansion (with his ex-wife, I think?), agreed to meet us at 7am and haul us, our boats, and gear back down to the boat ramp. He was super nice. There were several pickup trucks with boat trailers at the ramp. Pretty busy for 7am. Folks were heading out to go fish for catfish on this chilly Saturday morning. It was 45 degrees when we pushed off. I wore shorts and a windbreaker over my short-sleeve button shirt. Jeff had on about seven layers as usual.

Paddling back out the Yazoo Chute was fun and fast. What took us an hour and a half the other day took about 20 minutes this morning.

The Mississippi felt big to me today. Lots and lots of huge barges.

A pretty consistent wind out of the south the blew into our faces. The wind started at 5mph but was blowing 10-15mphs by midday. So, there was some wind chop, some current, some boat waves, and the river just felt really big to me. As in, I felt a little scared and vulnerable paddling my tiny 13-foot canoe.

Twice we paddled around a narrow 90-or-more-degree curve in the river, and both times, there were three to four huge barges lined up to pass through right as we got there. Not a huge deal, but it added some waves and tenseness to those sections.

I was feeling a little dejected today too, knowing that we need to get 30-plus miles in every day for the next nine straight days to be done by next Friday. Today was windy and slow going, which didn't help. I called Leslie and Seth, but neither answered. I was hoping for a little encouragement.

So, I resorted to music. I downloaded dozens of my favorite artists and albums over the past two days in Vicksburg. So, I listed to an entire James Taylor album, followed by a Patty Griffin album, followed by a Keith Skooglund album, followed by Rush's first album. That got me through three hours of paddling and helped to lift my spirits.

We paddled past a few houses that were sitting up on tall wooden stilts and pilings on the Louisiana side of the river. It's unusual to see houses along the riverbanks this far south, where the river rises and floods regularly. At one house in particular, it seemed like I just couldn't paddle past it. I'd paddle for 15 minutes and look to the right, and there it was. Then another 10 minutes and I'd look over and it was still to my right. And then another 10 minutes. . . . I finally said to Jeff, "I can't get past this damn house. Every time I look over, it's still there."

Jeff said, "So stop looking."

About 10 minutes later I snuck another glance . . . and it was still there!!

We stopped about mile 18 and got out of the boats to stretch and eat some lunch. And to let two big tows pass. Same old peanut butter and jelly bagel. But it felt good to take a break. It was probably in the low 50s when we stared out today, so my hands and wet toes were actually chilly, paddling in a breeze for the first few hours this morning.

My thoughts are turning more frequently to home with only eight paddling days after today. Although the pressure of having to average 31 miles every single day makes me feel better about being willing to take nine more days if we need an extra day. Sort of takes the pressure off.

For the last 10 miles of the day, we could see a big cooling tower for a nuclear power plant off in the distance. We camped on a big open beach, and the nuclear cooling tower is right across the river.

We had our final flag sticker ceremony this evening, adding stickers of the Mississippi and Louisiana flags to the side of our boats.

It's only 6:40pm, but the sun is already down. It's getting chilly and breezy. Jeff has a nice fire going. And our plastic whiskey glasses have some soothing brown liquid in them. I made dinner tonight, so it was Jeff's turn to collect firewood and make a fire.

Dinner was actually pretty good, relatively speaking. A small, canned ham that I diced up, added to some pre-cooked red beans and rice, and a package of precooked black beans. And then some hot pepper flakes dusted on top. I also had a huge dill pickle that I sliced up, and part of a chocolate bar for dessert.

Tomorrow, our goal is again to paddle 30 to 31 miles. Just keep taking it one day at a time, and the rest will just take care of itself.

It's supposed to be in the mid-80s tomorrow, so hot and sunny.

We have a nice beach campsite. A nice campfire. Full bellies.

The hardships of the day, the fears, the sore shoulders, have all fallen away, with the help of Advil and whiskey. Just enjoying the evening. Enjoying the opportunity to be camping on a beach along the Mississippi River (across from a nuclear power plant).

Boat Ramp? What Boat Ramp?

October 18, 2020 - Day 82
Mile 408 to Mile 382 - 26 miles
Total Mileage: 1,951

We're sitting in our collapsible camp chairs, eating smoked turkey sandwiches, barbecue chips, and dill pickles. See? This Mississippi River paddling is awesome if it weren't for all that damn paddling!!

Thirty miles per day is our goal to finish this river by Sunday, October 25th. Looking back over my journal entries over the past several days, I'm obviously obsessed about being done. We fell a little short today. There was a consistent breeze out of the southwest and into our faces for most of the day. And there were enough barges to keep us out of the main channel. Both combined to slow us down a bit.

It would be so nice to get a decent wind at our back, combined with a decent current, at least one day so we could go 35 miles and get back on track. But that's probably too much to ask.

It's hard to not obsess about finishing this trip. I've paddled over 2,000 miles over 81 days!! So, the last 215 miles and seven days seems so close to the end because it is so close to the end.

Temps hit the low 80s today, but the breeze helped. And during the last hour of paddling, we enjoyed some clouds, which was awesome.

Around mile 23, we rounded a sharp corner and met two large barges. One was a 35-er. It was at a very narrow point in the river, and the water close to the right shore, where we were paddling, was flowing backwards due to some massive eddy. And it was moving fast! And where this backwards flow met the main current, there were countless other smaller eddies swirling around and grabbing my bow and stern and jerking me in unpredictable directions. We had to paddle right through this mess to stay far enough away from the passing barges.

So, it was a lot of swirling water, big eddies, and paddling like heck just to stay in place with my bow heading downriver.

The highlight of the day was that we stopped at a boat ramp at mile 12 next to an abandoned Bunge company grain elevator, and then walked a mile to Fish Tales Café. It was a little café and grocery store in the middle of nowhere, just off a small recreational lake. We'd decided to stop just for something fun to do to break up the day.

We each bought huge hamburgers and salads for lunch. And Jeff got seasoned fries that we shared.

We also bought some pre-made smoked turkey sandwiches for dinner tonight. So, we took an hour and a half off the river, had a nice lunch, and made the day just a little more enjoyable. I put "boat ramp" in quotes because when we got to the spot noted on our map, the "ramp" looked like it had been washed away many years ago. No ramp in sight.

Boat ramps used to be destination points for us. Things to look forward to along the route because we knew we could pull our boats out, stop for lunch, and maybe even set up camp. Sometimes they even included a picnic table or an outhouse, or some nice green grass. But the last few have been total busts. All washed away and long gone. We passed by one this morning and never even saw it. Major floods must just wipe them out. The Army Corps maps need to be updated.

Our camp spot tonight is on a little sand spit sticking out into the river. We have great views both up and down the river. The wind has calmed. Jeff has a great beach fire going. Our smoked turkey sandwich dinner is over, with leftovers for tomorrow.

Another day of paddling is in the books.

Only seven or eight more days!!

We're going to shoot for 31 miles tomorrow and see what happens.

I just saw football scores on my phone, and the Packers got thumped by Tampa Bay and their "new" quarterback Tom Brady. I can't stand him.

I'm trying to do some birthday planning for Leslie's birthday in eleven days on October 29th. Not easy to do from the river.

Brother Dave and Steven

October 19, 2020 - Day 83
Mile 382 to Mile 350 - 32 miles
Total Mileage: 1,983

Laying in my tent at 7pm!!

Sweat is literally pooling beneath me and dripping off my nose as I lay on my stomach to write. I hate being sweaty and sticky in bed! And I double-hate being in my tent at 7pm!! WTF?

The mosquitoes where we are camping are horrendous. Despite spraying on several layers of bug repellent, and having a fire, the mosquitoes drove us into our tents way too early. It could be a long night!!

We woke this morning to a lot of humidity. We were both getting sweaty just packing up our boats. Jeff was up at 6am, out of his tent, and carrying dry bags to his boat by 6:45am. And we were on the water by 7:50am!

Once we started paddling, a slight breeze from the south felt great. And it was cloudy until almost noon, so it didn't feel too hot. The first couple of hours of paddling were nice. Perfect conditions.

It was 18 miles to Natchez. I talked with my mom, listened to music, and at one point asked Jeff to tell me a story. All to pass the time.

We paddled up to the impossible-to-miss boat ramp in Natchez just after noon. Natchez is well-known for its massive mansions that were built before the Civil War.

When we pulled up, there was a guy fishing off the end of the ramp next to his rusty beat-up pickup truck. He asked about our trip and said he'd keep an eye on our boats while we went up for lunch. At the top of the ramp are four or five old red brick buildings. Two restaurants, the old Mark Twain bar and hotel, and a gift shop. A couple of people over the past few days told us that most through-paddles spend the night at the Mark Twain, but we want to get more miles in today. We at least poked our heads into the cool old, iconic bar. It's hard for me to stop this early in the day. I get bored and feel like I'm wasting time. And stopping earlier in the day just means adding more days to the overall trip. On a one- to two-week paddle, it makes sense, but on a 90-plus-day paddle, I'd usually rather just keep going.

For lunch, we got tacos and I got a salad and a lemonade slushy. We also ordered quesadillas and chicken wings to-go for dinner tonight. When our food came out, the manager said that Dave Wunrow had just called in to pay for our meal! What? Apparently, Jeff had posted where we were on Facebook when we got to the restaurant. Jeff's brother Dave saw the post, and Dave called the restaurant and bought our lunch (and dinner). So nice! I don't often realize that there are family and a few friends who know we are on this trip.

It was great to stop in Natchez. Most of the huge mansions sit up on a high bluff overlooking the river. We didn't go into the town itself, but it'd be fun to check it out someday.

Back on the water at 1:30pm, it had gotten much hotter during our stop. Steven had done a good job of watching over our boats, and he offered us four ice-cold bottles of water from his Igloo cooler as we left. We also saw two heavily armed and bullet-proof-vest wearing "wildlife officers" talking with Steven as we got back to our boats.

The officers couldn't believe that we were paddling these tiny boats on this massive river, and several times told us to be careful, and to call them if we got into any kind of emergency or trouble on the river.

We paddled off into waves and oncoming barges. As we paddled beyond Natchez, the river seemed extra wide, and busy, and wavey. I felt uncomfortable sitting in my little open canoe, almost indiscernible on this gigantic, powerful river. The wildlife officers knew things about this river and what it can do to boaters. It took me an hour or so to settle down and settle into the rhythm I'd had this morning.

Our goal was to go 31 miles and the last few seemed to take forever. We finally pulled off after 32 miles at a spot that looked okay for camping. It ended up being a bit of a haul to find a spot that was flat enough to set up our tents. And, as I already wrote, the mosquitoes and the combined temperature and humidity are nasty. It still isn't really cooling off, and the sun has been down for an hour. We started a fire to burn our garbage, and had held out hope of staying up, but the bugs drove us in.

During our quesadilla and hot wings dinner, we reviewed our maps like we do every single night and made our plan for tomorrow. There is a boat ramp at mile 27, which could mean something, or nothing. And it's about 59 miles to the town of Simmesport and a motel. So, Simmesport is our goal for Wednesday night, for sure. We also cut off the Mississippi River in 47 miles and transition to the Atchafalaya, so we're really looking forward to that transition.

Tomorrow will be a long, hot paddle, with hopefully a few dips in the river along the way. And Wednesday should be much more interesting.

We're moving along!

Even the miserable conditions are all part of the trip.

Part of the adventure.

Part of the story.

But yuck! It feels nasty in this tent tonight!!

Tree Stumps Don't Stink Like That
October 20, 2020 - Day 84
Mile 350 to Mile 319 - 31 miles
Total Mileage: 2,014

Another good day!

We got in the miles that we needed to get in and have a couple of contingency plans for the next few days. Either we'll finish late in the day on Sunday (six days from now), or early afternoon on Monday. So, five or six more days to get to the Gulf of Mexico!!

Last night's camp sucked! Muggy. Buggy. Hot. Sticky.

I finally was able to pull my sleeping bag over me around 4am. Until then, it was just too hot. Jeff was up and packing by 6am.

Packing up was a muggy and sandy affair. That's the downside of camping on a sandy beach. Everything is wet from the dew and sandy. The tent fly was just dripping with water, and then gets stuffed in the stuff sack only to be pulled back out at the end of the day and set up, completely soaking wet.

Coffee and oatmeal while standing. My camp chair was wet too. We couldn't wait to get out of this nasty camp spot and get in our boats and start moving down the river. It was just a bad camp. Period.

It was cloudy overhead with some fog sitting on the water, so the air felt cool this morning. And the water was like glass, a rare sight on this big river. Paddling was really relaxing for the first few hours. It felt like paddling in the Boundary Waters in Northern Minnesota. I enjoyed it.

The clouds hung on until around 11am and then the sky cleared, and the sun came out, and it was like paddling in a blast furnace for the next four hours. Mid-80s, sunny, and no breeze.

We pulled off the river to stretch and eat some lunch on a sand bar around mile 20. After lunch, I was motivated to paddle hard, listening to Black Sabbath's Sabbath Bloody Sabbath album, and I ended up a mile or so ahead of Jeff. So, I pulled over and took a quick dip in the cool river. I probably dumped four or five coffee cups of river water on my head and dipped my Packer hat in the river several more times to cool off throughout the afternoon.

We're camped on a nice, level spot in the trees up on a bluff overlooking the river. We had to haul our bags up a steep sandy bank to get up to the flat part. I'd been excited about the thought of taking a dip in the river at the end of our paddling day today, so as soon as we'd hauled our gear up the bank and set up camp, I stripped down to my boxers and headed back to the river to rinse some sweat and grime off, and just cool down a bit.

As Jeff and I got down to the water, the ensuing dialogue went something like this:

Jeff: What's that big thing floating in the water?

Jon: That big brown thing twenty yards away?

Jeff: Yeah.

Jon: That's just a big floating tree stump.

Jeff: But it's shiny. Tree stumps aren't shiny like that. I think it's a dead animal.

Jon: That's not a dead animal.

Jeff: Yes. It's all shiny and bloated. And I think I can smell it.

Jon: Oh, shit. That's a bloated dead deer!

Jeff: That is a bloated dead horse. It's huge. I think it's floating towards us.

Jon: That's not a dead horse. It's not big enough. I think it's a bloated dead deer.

Jeff: Does it really matter?

So, our much-anticipated river swim quickly turned into a quick bird bath.

It's not as hot as yesterday evening, so sleeping should be more comfortable. But the mosquitoes are horrible!! We have a fire, and are both completely doused in bug dope, and they are still landing and biting.

It's so weird that it is hot and muggy here, and it just snowed six inches in St. Paul. I think we are going to be dealing with this heat and humidity and mosquitoes for the rest of the trip.

Tomorrow we lock through to the Atchafalaya River. It's been several weeks since we've seen a lock and dam. That'll be fun. We have several mile markers to watch for tomorrow to keep us motivated. And if all goes well, a motel tomorrow night! Which means showers and hopefully laundry! Sort of pathetic to reread that last sentence. But as my dad said often during his last few months on this earth, "Who cares?"

Hello, Atchafalaya

October 21, 2020 - Day 85
Mile 319 to Simmesport, Louisiana - 27 miles
Total Mileage: 2,041

Jeff was up again at 6am, packing up in the dark.

Our campfire was cut short again last night due to the fucking mosquitoes. It wasn't as hot and sticky in the tent as it was the night before, so reading myself to sleep was okay.

I started packing up the inside of my tent by headlamp around 6:30am. Jeff was already down at his kayak, organizing his bags, and had his stove going, heating water for coffee. The bugs were out and forced us into our boats and on the water by 7:40am.

The sun was hot by 9am. Not a cloud in the sky. And not a breath of wind. I was dumping cups of river water on my head to help cool off by 9am!!

We were both really looking forward to today. We had several notable mile markers to paddle towards throughout the day. Things to look forward to. Mile 3: First overflow channel to Red River. Mile 5: Second overflow channel. Mile 7: Third overflow channel (can't you feel the excitement building). These overflow channels are used when the Mississippi River gets too high, so excess water can be diverted into an adjacent river. They are basically huge locks, that we've been warned to stay on the opposite side of the river from, especially if we hear a loud siren announcing the opening of the locks.

To continue with the exciting mile markers for the day . . . Mile 14: "River Ferry" parking lot (a possible rest stop that we didn't stop to rest at). Mile 16: Turn off to the lock and dam that leads to the Atchafalaya River.

Even though we'd been watching carefully for the right turn into the channel that led to the lock and dam that leads into the Red River and then to the Atchafalaya, we both paddled right past it, and didn't realize it. You can't see the lock and dam at the turn off, and it just looked like a dead-end to me. Plus, the river was moving right along, so we zipped past pretty quickly. Once we realized our mistake, we had to paddle back against the current to get into the channel.

The heat and sun were oppressive today at 86 degrees, with a heat index in the 90s. I dumped cup after cup of river water on my dead and down my back. It only cooled me off for a few seconds but felt really good for those few seconds.

We were met at the entrance to the lock by three loud horn blasts. Jeff had called ahead about 45 minutes earlier to let the lockmaster know we were coming. The lock seemed small and narrow compared to the ones we've been paddling through. It allows tows that are only one barge wide and maybe two or three long to enter. And I don't think this lock sees much commercial or recreational boat traffic. We were definitely the only boats around.

The lockmaster peered down at us as we paddled into the lock and asked some questions about our trip. He told us to tie off on the "floats" positioned along the concrete wall of the lock, which was something new for us after the 29 locks in the Upper Mississippi, where we just held on to ropes that were hanging down along the lock walls.

The drop to get to the river level of the Red River was at least 18 feet, after which we paddled out into a two-mile channel that led to a four-mile section of the Red River, before reaching the Atchafalaya. Those six miles in the channel and Red River were flowing backwards for some reason. Only 1-2mph, but we had to paddle hard and non-stop in the unbearable sun and heat just to poke along at 2mph. It was a long, hot three hours through the narrow waterway that was lined with tied off barges along both sides. None of the barges were moving, and we didn't really see anyone along the banks. It seemed like it would go on forever, and we both were getting dejected and irritable.

After a few hours of struggling, we entered the Atchafalaya! Finally!!

There was a slight current for our six-mile paddle to Simmesport, so it was a welcome relief from the last few hours of struggle.

As we neared and then crossed underneath a train trestle and road bridge, a local guy in a flat-bottomed skiff zipped over and asked if we were from around here. He told us to stay out of the middle of the river (where we were) because beneath the bridge, there were huge eddies that would likely flip us over.

He said, “Be careful. It’s dangerous.”

Well, we adhered to his words of caution, and paddled more towards shore, but as we neared and passed beneath the bridges, we saw no current, no eddies, no waves. Nothing. So, who knows what this guy was talking about? Maybe when the water level is higher?

As we passed under the bridges, I called a motel in Simmesport to see if we could get a ride, since the motel was a few miles off the river. I’ve called them each of the past two days, saying we were coming, and hoping for a ride.

Jay, a young guy, and the apparent owner of the motel, met us in a brand-new minivan. We quickly realized that he wouldn’t be able to haul our boats, and Jay asked me if I knew how to drive.

Sure.

So, I hoped in the passenger side of the van, and he drove me to the motel and gave me the keys to a 20-foot-long U-Haul truck, which I proceeded to drive back to the landing. We loaded both of our boats and gear into the U-Haul. Perfect. And he said we could just keep our stuff locked in the truck overnight. Super nice.

Juxtaposed to Jay’s over-the-top help, The motel was a complete and utter shithole. Probably the grossest motel I’ve ever been in in the United States. The first room that he gave us a key to literally had filthy pillows and bedding that looked slept in. The two rooms we finally settled on had no bath soap, no towels, filthy walls and carpeting, and Jeff’s room had no lamp or ceiling light. Just a bathroom light.

So, we are both sleeping in our sleeping bags on top of the

beds tonight.

Super nice guys, Jay and Ray (probably not their real names?), but a filthy motel.

Jay first drove us to a local laundromat that turned out to be closed down. He agreed to do a load of laundry for us at his house, wherever that was. Then he dropped us off at a seafood restaurant about a mile away. Rabalais Seafood. It was a little mom-and-pop place, literally. We walked in, and the owner, Ray, was sitting at a table peeling a mountain of fresh shrimp.

We ordered seafood platters of shrimp, clams, catfish, frog legs, and slices of white Wonder Bread. The food was fresh and tasty, and we washed it down with a few cold Michelob beers.

Fox News was blaring on the huge wall-mounted TV while Ray made occasional jokes about Obama and Biden. We just ignored them and did some internet stuff while staying as long as we could to charge our devices (since most of the outlets in the motel didn't work) and waiting for dinner.

Ray brought out coffee after dinner and asked about our trip. He was funny and friendly, and very sarcastic. He looked and acted like my friend from Sitka, Alaska, Jeff Budd. Ray could've been Jeff's brother, they looked so much alike.

Ray offered us a ride back to the motel. His car smelled like weed, and his dog "Shithead" jumped in with us.

Jeff is convinced that the little mini mart next to the motel, also apparently owned by Jay and Ray, is a major drug dealing spot. There is a constant flow of people driving up in fancy cars, not going into the store, going to the rolled down window of the next car over, and then getting back in their car and driving off. The owners definitely do NOT seem to be making their money on renting motel rooms, or even on the mini mart.

Regardless, Ray and Jay have gone out of their way to treat us nice, so that's all that matters!!

Tomorrow is going to be another hot one, but we've decided we are going to have to paddle some next Monday, adding a day to our trip, so we are only going to paddle 25 miles tomorrow and stay with the "famous" Atchafalaya River Angel Danny Majors.
What a great decision that turned out to be!

Danny Major, Saint of the Atchafalaya
October 22, 2020 - Day 86
Simmesport, Louisiana, to Danny Major's (Mile 30 on the Atchafalaya River) - 24 miles
Total Mileage: 2,065

Danny Majors.

That's all that needs to be said.

What a trip.

Danny is all about helping other people. That's what he does. Every paddler that comes down the Atchafalaya stops to stay with Danny. For a day or two, or longer. If I didn't want to get home so bad, I'd vote to stay at least another day for sure. Danny opens his home to literally anyone that happens by. Folks interested in his organic farming, couch surfers, paddlers. Anyone. We've been hearing about Danny Majors from several people for weeks, and now we know why.

Danny is a disheveled, funny, trusting, people-loving, trying-to-do-a-dozen-things-at-once, offer-you-the-keys-to-his-car, kind of guy, who is full of great stories.

As we visited through the afternoon, Danny shared about his life, and his two kids, and the scores of people who have stayed at his house over the years. While he talked, he was also making an amazing red beans, sausage, and rice dinner. Danny loves to cook. He decided at the last minute to make potato salad because he had a ton of potatoes. Danny said he'd never made potato salad before, but he threw together the ingredients that he thought seemed right, and it was amazing.

We arrived around 3:30pm and sat up talking until well after 10pm. Super late for us. I didn't get any computer work done, and around 11pm, I'm just starting to write in my journal.

"I'm left-handed."

"I buy all of my clothes at Goodwill."

"I don't paddle a canoe or kayak. But I'm intrigued by people who do."

"My kids are my best friends."

Those were Danny's answers, when I asked him to tell me about himself.

He spent the last ten minutes telling us that we need to stop in the next town downriver, Krotz Springs, and that we have to walk into town and go to Billy's and buy some "boudin" and "boudin balls." Danny says it's addictive, "like crack." He said we absolutely have to stop. So, as soon as we get out of here in the morning, we'll paddle 10 miles, and make a pit stop at Krotz Springs.

Danny has about 20 cows, 10 guinea hens, a Shetland pony, a big white dog named Toga, and a farm cat named Meow.

I could write and write and write about Danny Majors. Every

time he opens his mouth, a great story comes out. Maybe I'll write more later.

He lives out here all by himself. His kids come to spend the summer. He's a little zany.

Danny follows all of the Mississippi River through-paddlers online. So, he has either met or knows all about lots of the paddlers we've heard about or read about. He's totally tuned into the Lower River paddlers, and knows all their stories, but he's never paddled. He has no interest in paddling. He does have a shrimp boat and goes shrimping.

Story, after story, after story.

Danny has taken so many people in over the years. Some for an overnight, like us, and some for weeks and even several months. He has a very interesting couple staying in his cottage next door for an indefinite period of time. He took them in because they were down and out, and they just keep staying. He usually puts paddlers up in that cottage, but since its occupied, I'm staying in Danny's kid's bedroom and Jeff is sleeping on the recliner.

Okay, time for bed. We need to get up and paddle tomorrow and get some of that boudin at mile 10.

Danny gave us a little history lesson on Cajun people and their origins. He said this entire area of Deep South Louisiana is called Acadiana. The Acadian people, originally from France, emigrated to the U.S. and were pushed into Nova Scotia, Canada, by the British. I remember when Leslie and I did our three-week bicycle trip along the southern coast of Nova Scotia, we met lots of Acadian people and passed through some communities where everyone still spoke French. Some of the French Acadian people ended up leaving Nova Scotia and going down to New Orleans and the surrounding areas.

The word "Cajun" is an adaptation of "Acadian." When you say "Acadian" fast, it sort of comes out "Cajun." Danny said that Cajun food was never a special gourmet thing. It just resulted from Acadian people making use of what they could get their hands on, namely rice, beans, and seafood. We are deep in Cajun country right now.

I forgot to write about a guy we met the day we stopped in Natchez for lunch. When we pulled our boats out at the Natchez boat ramp, an older scrawny guy walked down to meet us. He said he'd been on the river for a year and a half, that his boat was parked "somewhere down the river" and that he'd "taken his bike up to Natchez to stay for a while." He said over 200 articles had been written about him. He seemed a little wild-eyed to me.

Well, we were telling this story to Danny, and he asked, "Was his name Kelly?"

Yup.

Danny showed us a photo of Kelly, and that was him.

Danny said that people up and down the river have been talking on Facebook and paddling blogs about Kelly for the past year. "He tells a very hard-times, down-on-his-luck story. Cancer diagnosis. Has all of his earthly belongings with him on his boat. Overstays his welcome wherever he goes."

And I said, "Wait a minute. Did someone further up the river give Kelly a boat, because his had somehow gotten a hole in it?"

"Yeah, man. That's exactly right," Danny piped in.

Jeff and I both said almost simultaneously, "That's what Paula and Todd were telling us about a while back. A guy they met on the river was telling them stories of one calamity after another, and Todd gave him one of his boats."

So crazy! The small world of Mississippi River travelers . . . how all these people we've met, have met or heard about other people we've met or heard about. This river community, and paddling community, is a very small group of interconnected people. Very cool.

Danny also knew all about Rebecca and had heard several Rebecca stories from other paddlers and river bloggers. Crazy.

Ray's the Man

October 23, 2020 - Day 87
Danny Major's to Hwy 10 Bridge - 30 miles
Total Mileage: 2,095

It's raining hard outside. Our first real rain in 41 days, which is pretty wild. We'd just finished dinner and the mosquitoes were driving us into our tents by 6:30pm (bummer), and thunder and lightning started in, and now it's been raining pretty good and hard for the past hour.

Having to go into our tents at 6:30pm blows. But it could be worse. It's warm and muggy again tonight. And with the tent flaps closed because of the rain, it's pretty damn sticky in here. I'm sticking to my Therm-a-Rest as I lay here writing about the day.

Today was a long and psychologically challenging paddling day. Just one of those days where we paddled hard and non-stop pretty much all day. And the miles just didn't pass quickly enough. The conditions actually weren't bad. Mostly overcast and almost no breeze. And really flat water. Low 80s and muggy, but not as bad as the past couple of days.

Just sort of demoralizing.

My body seems to be breaking down.

My shoulders start aching earlier in the day, and ache longer after I stop paddling.

My butt aches more.

I have to reposition myself on my canoe seat, sometimes several times an hour, just to help with the aching.

I called Leslie around 3:30pm, mile 24 and a half, and that really helped. We talked and talked about Danny, and Kelly, and Rebecca, and other river characters. And before I knew it, we were approaching the Highway 10 bridge and mile 29 for the day. It was awesomely helpful.

For dinner, I made a really great mac and cheese with seasoned panko flakes that Leslie gave me a while back. We added three cans of tuna from Ray (more about Ray in a bit) and it turned out great!!

Did I mention that Danny made us a pot of coffee, bacon, eggs, and buttermilk biscuits for breakfast this morning? So awesome!! And then we sat around and talked and drank coffee . . . which is why we weren't on the water until after 9am this morning.

Oh, and we never stopped to buy boudin at mile 10. By the time we got there, the paddling was slow, and there really wasn't a good place to land our boats, and we both were still full of bacon and eggs, anyway.

But the highlight of the day, and maybe of the entire trip, was meeting Ray at the Atchafalaya Campground.

Around mile 16, Jeff and I were both feeling exhausted. We spotted a little campground up ahead that wasn't on our map. It was pretty unusual to see a campground that was this close to the river. There was a nice concrete boat ramp, so we pulled off, tied up our boats, and lugged our lunch stuff and water bottles up the ramp to a little, very shabby covered picnic area.

The campground was actually a very run-down trailer park. There were maybe five or six campers and a big yellow school bus that looked like they were all pretty permanent, and that was it. The cinder block restrooms didn't have doors, or paper hand towels, or even toilet paper.

Sitting under the covered picnic shelter in a wheelchair was an older guy named Ray. We asked if we could sit down, and Ray welcomed us with open arms. He proceeded to ask us all kinds of questions and could not believe that we'd paddled 2,000 miles over 87 days.

Ray immediately got very protective, and told us, "Whatever you do, don't camp on the other side of the river over there." He pointed. "There's feral hogs over there."

Okay.

Ray also told us to be careful of alligators and that we'd see a good size one about a mile and a half downriver. (We never saw it.)

As soon as we sat down, Ray said, "Let me get you both some ice-cold water."

So, he got out of his wheelchair and slowly walked a few hundred yards to his camper and brought us back four frozen bottles of water. So awesome! We were so grateful.

Ray told us that he had to move his trailer out of a different trailer park near Baton Rouge because the monthly rate went up to $450 per month and he couldn't afford it.

Ray lives permanently in his little travel trailer.

Ray doesn't have much.

As we finished our lunch and prepared to leave, Ray said, "Now, y'all don't have to leave yet, do you?" We told him we still had 14 more miles to paddle. Ray said, "Let me get you some more ice water and some Ritz crackers."

I said, "Ray, you just keep sitting there. Don't worry about us. We're fine."

"No," he replied. "Don't leave yet, you need some more cold water out there in this hot sun." So, Ray got up out of his wheelchair one more time and walked all the way back to his trailer and came back with a plastic grocery bag of four more partially frozen water bottles, three cans of tuna, four Pop-Tarts, and a container of pepper.

Ray said, "I'm sure you can use this. I'm sorry I don't have more to give you. I don't have much."

I almost started to cry.

It was the biblical parable to the Widow's Mite playing out before our eyes. Ray giving us some of the little that he had. So incredibly awesome!

All afternoon, Jeff and I talked about Ray. And we had Ray's tuna fish with our dinner tonight, which made it especially tasty. It was a dinner of love. Leslie had brought us this meal, and Ray added the tuna.

As we were getting into our boats after lunch, Ray said, "So what do you think about this election? Who you gonna vote for?"

I said, "Well, Ray, if you want to know the truth, I'm voting for normalcy and decency. I'm voting for Joe Biden." I said, "Biden is like you, Ray. He wants to help people and bring them together."

Ray replied, "I've always been for Trump, but I agree with you. We need to come together." He continued, "It doesn't matter what color you are, we can all get along together and help each other out."

Ray. Made my day. Made my week. Made my trip.

Redneck Wild Pig Hunters and Where the Hell is Jeff?

October 24, 2020 - Day 88
Hwy 10 Bridge to Mile 90 - 30 miles
Total Mileage: 2,125

We are camped on the left side of the river in a little clearing in the woods. I feel lucky to be here. Jeff is kind of anxious and clearly irritated. We are camping at a spot that some locals cleared out for a wild pig hunting camp. The small clearing is about 75 feet away from two five-gallon buckets of molasses hung up in a tree that are slowly dripping their contents onto a trail of dried corn kernels. The corn kernel trail leads to a camouflaged metal cage for catching wild hogs. How the hell did we end up here? It's kind of a long story.

We started watching for a place to pull off and camp around 4pm. This part of the Atchafalaya doesn't have any sandbars or sandy beaches or grassy clearings along the river side. Just high and steep muddy banks on both sides. We pulled our boats up to a couple of possible landing spots, but one was too muddy to get out of the boats, and the other was too sloped and brushy to set up our tents.

After paddling another mile, we saw a couple of flat-bottomed skiffs with some guys standing on a small dock up ahead. There were a couple of small houses on each side of the river. We paddled up and asked them if they knew of a place to camp . . .

like the grassy lawn of the cabin they were standing in front of, for example.

They told us they had a "camp" just downriver, with a clearing and a metal table. And they pointed to a little peninsula a ways down the river on the opposite side. They were all drinking what looked like cold, cheap beer, and they had a Louisiana accent that I had a hard time understanding. There were six of them, and one guy was especially loud and especially drunk.

As we paddled away, Jeff said that he didn't feel safe, and was imagining them killing us in some gruesome manner.

The loud drunk guy jumped in his boat with another guy that looked like an LSU linebacker, and they sped out ahead to apparently show us the way to their camp. When we caught up to them about 15 minutes later, they were smoking pot and drinking and yelling. None of which helped build Jeff's confidence in the situation.

These two guys showed us where to pull up and they jumped out of their boats. Shit.

The shore was super muddy and immediately swallowed up our feet to our calves, which made unloading the boats an added frustration. The two guys eagerly led us over to the system they had set up to lure the wild hogs into the metal trap. At one point, the loud guy offered me his joint. No, thank you.

And then he turned to Jeff and said, "You're in man's country now. I know you are packing a gun for the wild animals. So, if you hear a wild hog or a sheep just shoot 'em."

"You got oil for cooking, don't you?" (And "oil" came out like "aawll.") "Well, just butcher up that hog, eat what you can, and throw the rest in the river."

Jeff just went along with it all. The whole scene was totally freaking him out, he told me shortly after. Jeff told me earlier today that he is super nervous about alligators, and then we saw a bear a couple of days ago, and now wild hogs, and two drunk, stoned Louisiana bijou boys. It was all just a little too much for Jeffie. I was just hoping the guys would leave soon so we could set up our tents, eat, and go to bed.

As we unpacked our boats, and schlepped our bags through the mud, Jeff said, "This is the worst campsite we've had on this trip."

I said, "Really?" We had a little clearing in the trees for our tents, a slight breeze, no sand to penetrate everything we owned, and it wasn't raining. I was totally happy we'd found a spot, and, as usual, just glad we were done paddling for the day.

And these two potential good-ol' boy murderers left us with five ice-cold bottles of water and five ice-cold flavored alcohol tonic waters along with an almost empty spray can of Off for the mosquitoes. Super nice. They told us to stay safe and wished us luck. And then they sped off back to where we'd first met them.

We passed a dozen or so little shacks along the river today. No electric lines or roads. So, only accessible by boat. Otherwise, we paddled through 30 miles of nothing since leaving the Interstate 10 bridge this morning.

It rained last night until 9pm or so, and it stayed warm and humid all night. I didn't actually get into my sleeping bag until several hours later. Packing up this morning, everything was sandy and wet. Even my sleeping pad seemed like it had soaked up the humidity.

Tonight, as I write (we were driven into our tents at 6:50pm by the mosquitoes), it feels a little cooler and not as muggy.

The big drama of the day, and probably why Jeff has seemed extra agitated this evening, went down like this. . . .

We'd stopped on a little beach around mile 16, around 12:30pm. Jeff was back in his boat first and floated off in the slight current. I stayed back to pee before heading out behind him. As I paddled off, I moved into the middle of the river to try to find faster current. Jeff was just ahead, and to my right, paddling closer to the shore.

After about ten minutes, I saw a small barge coming upriver towards us, so I paddled to the left side of the river, opposite Jeff. It took the barge 10 minutes to pass, as I paddled along the far-left bank. I could see Jeff over on the far-right side, now a little behind me. About 10 minutes later, I looked across and backwards, and saw two skiffs heading towards Jeff, so I figured they'd seen him, and wanted to ask some questions about what he was doing and where he was heading. So, I just kept paddling on.

As a second barge came into view, heading upriver, I decided to cross back to the side Jeff was on. I glanced back and didn't see his boat, and figured he was still back there somewhere, talking with the guys in the skiffs.

Maybe 10 minutes further downriver, I turned my boat around, and scanned for Jeff.

Nothing.

Distances are deceiving on the river. And it's easy for our little boats to blend into the surroundings. Especially if he was still a ways back.

No sign of Jeff.

I moved back towards the middle of the river and just bobbed around for a while, hoping it would make it easier for Jeff to spot me. I waited for about 15 minutes with the slight current carrying me along. Still no sign of Jeff. I wasn't even totally sure if he was behind me or in front of me?

At this point, I realized that he'd lost sight of me, and probably never even knew that I'd been paddling on the other side of the river instead of directly behind him, and now he was paddling back to where we'd stopped for lunch? Maybe?

The wind and current were too much for me to want to start paddling back upriver, so I decided to just stay put in the middle, easier to spot, and wait.

I started thinking through my options. How far should I float or paddle before stopping to set up camp? I tried calling and texting several times, but there was no signal.

I finally decided that the best thing to do was just stay out in the middle of the channel. Not paddle. And flag down a boat if another one came along. It was a Saturday, and we'd seen five or six skiffs today.

After about 30 minutes, a little green skiff came ripping downriver towards me. I waved him down and asked if he'd seen a kayak.

"Yup. About a mile back," he said.

I asked him if he'd go back and tell Jeff that I was up ahead, and okay.

"No problem."

He had some insulation and a stack of plywood on his boat, probably heading out to fix up his river shack.

About 15 minutes later, the guy in the green boat was back, and said he'd found Jeff. And, it turns out, Jeff had been looking for me and had started paddling back upriver to where we'd stopped for lunch. So, I just floated for another twenty minutes or so until Jeff caught up.

Jeff had been really worried that I'd either been run over by that first barge or had a heart attack and died at our lunch spot. I'm not totally sure where that option came from? But he was seriously worried.

In thinking through his options, he couldn't imagine that I'd somehow gotten ahead of him, so he was sure that something had happened. We were both super relieved to be back together. It was a little traumatic for both of us. And not having any cell signal, or way to communicate, complicated the situation, especially if something had happened to one of us.

Just now, Jeff yelled over from his tent, "Jon, I'm glad you're not dead."

I'm glad we found each other, and that I'm not camped here alone tonight, wondering where the fuck Jeff is.

Two more days.

So hard to believe that we are this close to the gulf. Maybe 46 miles to go?

We've paddled over 2,100 miles since Lake Itasca State Park. And over 1,000 miles on this second half of the trip. That's a long damn way!!

The End is Near

October 25, 2020 - Day 89

Mile 90 to Wax Lake Lodge - 25 miles

Total Mileage: 2,150

I started getting a little choked up towards the end of our paddle today. Thinking about this 89-day effort to paddle the Mississippi, the hundreds of thousands of paddle strokes, the scores of nights camping, all of the people we've met along the way that have helped us out and encouraged us, the heat, the aching shoulders, the scary moments, and the paddling, day after day after day.

It's a lot.

Leslie's sacrifices; Seth's support; my mom's daily calls. It's a lot.

I scoured the shore for alligators all day today.

Didn't see a single one.

It was actually a cool night last night. Great for sleeping. A welcome change.

We both woke to the sound of a boat motor. Some of the group we'd met yesterday pulled up in the dark and got dropped off. I heard one of the guys say, "It looks like they're still here." As they walked past our tents, I asked if they'd brought us anything from Starbucks.

No answer.

They walked off in the early morning darkness to their deer stand to do some hunting. I downed my oatmeal and coffee, and we packed up pretty quickly. We were on the water by 8am.

It was foggy and cool with a nice breeze coming out of the north, and at our backs. Perfect paddling conditions. It stayed foggy until around 11am. A couple of small tows passed us in the fog. I love that. It was cool to watch these boats appear and disappear in the fog ahead of us.

It was 14 miles on the foggy Atchafalaya to Yellow Lake. We paddled two miles along the west shore and then pulled off for lunch at a small boat ramp. More peanut butter and jelly on a bagel. I'm pretty sick of all my food. And have no idea why my food bags still weigh a total of about 50 pounds!

Yellow Lake was six or seven miles long, so it took a while to cross. The sun came out. The temps heated up, and it felt like it took forever to get to the Wax Lake cut-off that Park Neff had told us to watch for and shown us on the Google map projected onto his living room wall.

We followed the channel for four more miles to Wax Lake Lodge, with it's three single-wide trailers sitting up on cinder blocks for when the river floods. I'd contacted them for a place to stay tonight, but they were full. It's fun to actually see a spot that I've been looking at on a map, or online for weeks and weeks.

Travis, the owner, was out grading the parking lot with a bulldozer when we pulled up. Great guy. Super friendly. It would've been a great place to stay, right on the river. He brought Jeff and I out two ice-old Cokes and let us store our boats there for the night. We'll be back in the morning to finish this thing up.

Jeff's very good friend Shawn came to pick us up and take us to a hotel in Morgan City, about 10 miles away.

Mexican restaurant.

Margaritas.

It's 8:30pm. Well after my bedtime. Quite a different setting from where we camped last night.

One more day! Day 90!!

Crocodiles Rock

October 26, 2020 - Day 90
Wax Lake Lodge to the Gulf of Mexico - 21 miles
Total Mileage: 2,171

Phew!

Done!

In the books!

Relegated to memory and photos and my journal that no one will likely ever read.

It feels weird to be done.

I've been thinking about the Mississippi River for the past two years. Ever since my mom told me that my cousin Jeff was planning to paddle the river.

Now it's time to move on.

I'm glad that I got to share this trip with Jeff. One person in my life who understands what it's like to paddle 90 days, almost 2,200 miles. And to experience all that this river has to offer.

The river has so many facets, so many moods, so much personality.

Every day it offers something different. Some days, the river is tranquil and meditative, and some days it is violent and disturbed.

Some days it can be both, in the same day.

It was so fun to be paddling this last 20 miles of the river along waterways that we've been studying on Google maps and Google Earth for weeks.

We had originally planned to paddle through New Orleans and on out to the gulf, and Mile 0 on the Army Corps map. But early on, several people suggested that we consider the "Atchafalaya Finish" option. So, I asked Leslie to mail my Atchafalaya Army Corps maps to a friend of Mike's in Delta, Mississippi. Then, Park Neff told us about the Hog Bijou variation, and we were sold.

Instead of paddling the Atchafalaya River through busy and boat-clogged Morgan City (which we were able to see last night as Shawn drove us to our hotel), and then 25 miles out to the gulf, past all roads, and then needing to arrange for a fishing boat to come out to pick us up, we opted to turn off the Atchafalaya River at Six Mile Lake, a.k.a. Grand Lake, a.k.a. Yellow Bijou. Then, into the Wax Lake Channel, and along the Wax Lake Pass, across the Intercoastal Waterway that heads east and west, and then finally, to take a right onto the calm and winding Hog Bijou that drains into the Gulf of Mexico.

What a great decision! The first 10 miles on the straight-as-an-arrow Wax Lake Pass was really pleasant. A very slight current and a breeze at our backs helped us along on our final day. There were several beefy, welded aluminum fishing boats built for the ocean that zipped past us.

We were a little worried about finding the spot where Hog Bijou turned off. Park had given us some things to watch for, and Jeff had saved a satellite image on his phone. But Hog Bijou was the perfect way to end our adventure. It reminded us of the river during the first couple weeks of our trip heading out of Lake Itasca. Narrow, calm, and scenic. Lots of birds, and no other boats or people.

The only difference? Alligators!!

I've wanted to see an alligator for days now. And Hog Bijou delivered. We saw 67 alligators today! 67!!! They were everywhere.

Mostly we'd see their two eyes, and often their U-shaped snout sticking just above the water line. Watching us paddle by. Occasionally, I could see an entire body floating just beneath the surface. Some were pretty close, drifting just 10 or 15 feet away. And some were further away. But Hog Bijou is only about 75 feet wide, so none of the alligators we saw were too far away.

Every time one of us spotted an alligator, we'd call out the number "10" or "11" or "36" or "37." Jeff was super nervous initially about the whole alligator situation. A couple of days ago, he told me he was really worried about seeing them. But once we'd seen a dozen or so, he started feeling more comfortable, and we both were eager to see how high our alligator count would get. I thought it was awesome. Especially seeing so many. At one point, Jeff and an alligator got a little too close for each other's comfort, and the big alligator made a sudden move to dive underwater that surprised all of us.

The wind stayed at our backs all day.

As we neared the gulf, maybe the last mile, there was no current, and we started seeing lots of white egrets, blue herons, and other birds.

A few times, I thought about this great adventure coming to an end, and got choked up thinking of Leslie's support, Seth's constant encouragement, my mom's calls, and Monkey Face's daily presence, tied to the bow of my canoe. It all helped.

Jeff's buddy Shawn was waiting for us on a small channel that cut off from Hog Bijou. Jeff and I paddled past the channel another quarter mile, and out into the Gulf of Mexico. We wanted to paddle past all of the trees and stumps, so we could just see open water in front of us.

Done!

We took selfies. Took several photos of Monkey Face and Peaches (Jeff's paddling mascot). Jeff took a photo of me holding a copy of the Hickman County newspaper that I've been carrying along with me since Columbus, Kentucky.

And then I called Leslie, Seth, my mom, and Tyler to share the finish of this epic adventure with each of them as I bobbed around in the gulf.

What's next adventure-wise? Croatia with Leslie, a long-distance hike in Ireland with Leslie, hiking the Baekdu Daegon Trail in South Korea, a bicycle trip in Newfoundland, a motorcycle trip with Seth in Mongolia, climbing the highest peak in Peru, climbing Pico da Neblina in Brazil, another long paddling trip with Jeff, hiking the Appalachian Trail. . . .

Who knows?

I'll hopefully do all of them over the next several years.

There is always another adventure.

Always another trip to plan.

But right now, I need to go home and get re-acquainted with my wife and get my work life back on track.

And go celebrate my grandson Arlo's first birthday!!

There is nothing more important than that!

Back on the River in Clarksville, MO – Day 48

Happy to be Back to a Paddling Life

Irish Band in Alton, IL – Day 50

Portaging around a Small Falls Near Granite City, IL – Day 51

Too Many Barges Lined Up to Go Through the Lock, So We Went Around It – Day 51

Nearing St. Louis, MO – Day 52

Wrong Direction in Search of Something Cold to Drink – Day 53

End Result of Heading Up a Dead-End Stream

Welcome to St. Louis, MO – Day 54

Paddling Past the Arch

Pelicans

Oddly Ironic Sign on a Sunken Boat

Essentials

36 Barges Heading Our Way

Napping on Day 62

Raining and Cold on Day 63

Tortilla Lunch #4 – Honey, Peanut Butter, and Aunt Joanna's Snack Mix

Tortilla Lunch #5 - Processed Chicken and Spicy Blue Corn Chips

Tortilla Lunch #6 – Spicy Tuna and Wheat Thins

Memphis, TN – Day 66

Cold and Choppy Paddling

Welcome to the State of Mississippi – Day 67

Sign? What Sign? – Day 68

Dinner Time. My Turn to Cook.

Worth the Wait?

Greenville, MS – Day 76

Southern Hospitality – Day 77

More Southern Hospitality. Indulging in Hank's Hootch.

Heading to Vicksburg, MS – Day 80

Laundry Day – Day 82

Paddling Past a Barge – Day 83

Locking Through to the Atchafalaya River – Day 86

Relaxing at Danny Major's – Day 87

Paddling on the Atchafalaya – Day 90

It's Dead. But Still!

Entering Hog Bijou on Our Final Day – Day 91

Gators in Hog Bijou

The Gulf of Mexico – Day 91

Another Completed Adventure with Monkey Face

Brothers for Life

Celebrating the Finish

EPILOGUE

In 2019, my Adventure List didn't include any paddling trips. But now, at the beginning of 2022, it includes several, all with Cousin Jeff, of course:

- Paddle the 2,341-mile Missouri River – 14 weeks
- Paddle the 688-mile Cumberland River – 4 weeks
- Paddle the 1,980-mile Yukon River – 10 weeks
- Paddle the 1,469-mile Arkansas River – 9 weeks
- Paddle the 981-mile Ohio River – 6 weeks

Jeff and I finished our Mississippi River paddle in 2020.

In the Spring of 2021, we paddled the 652-mile Tennessee River, referred to as the "Tennessee River Line 652" from Knoxville, Tennessee, to Paducah, Kentucky. A book about that adventure will hopefully be out by the end of 2022.

In the Spring of 2022, Jeff and I were planning to paddle the first 1,000 miles or so of the 2,341-mile Missouri River that begins in Three Forks, Montana, and continues across the states of Montana, North Dakota, South Dakota, Nebraska, and Missouri before it enters the Mississippi River just north of St. Louis. But in the Fall of 2021, Jeff suffered a major stroke. Over the last couple months, Jeff has been spending every waking moment focusing on his rehab regime of physical therapy, occupational therapy, and speech therapy. Jeff is committed to getting back in his kayak, and eventually embarking on, and completing our Missouri River trip.

If anyone can come back from a major stroke and paddle a 2,000-plus-mile river, it's my cousin and paddling brother Jeff.

Author Jon Wunrow is a grandparent, parent, husband, adventurer, grant writer, cabin builder, beer brewer, Tribal advocate, and Green Bay Packer fanatic who occasionally finds time to plan and enjoy long-distance adventures around the world.

In addition to pursuing his passion for climbing many of the highest peaks in the Western Hemisphere, he has also hiked the 2,650-mile Pacific Crest Trail, paddled the Mississippi and Tennessee Rivers, climbed Kilimanjaro with his son Seth, hiked the 870-mile Wales Coast Path and the 630-mile Southwest Coast Path with his wife Leslie, and has enjoyed cycling, hiking, climbing, paddling, and beach sitting adventures in dozens of countries.

Wunrow is also the author of *Never Stop Walking: A Wale's Coast Path Adventure*, *Adventure Inward: A Risk Taker's Book of Quotes*, *High Points: A Climber's Guide to Central America*, and *High Points: A Climber's Guide to South America, Part 1.*

Printed in the USA
CPSIA information can be obtained
at www.ICGtesting.com
LVHW020343111224
798853LV00024B/441

* 9 7 8 1 7 3 6 3 8 7 0 4 7 *